another
Reason
to visit ü

Gone Fishin'

Ron Bern and Manny Luftglass

Rutgers University Press
New Brunswick, New Jersey, and London

Library of Congress Cataloging-in-Publication Data

Bern, Ronald Lawrence, 1936–
Gone fishin' : the 100 best spots in New York / Ron Bern and Manny Luftglass.
p. cm.
ISBN 0-8135-2745-7 (pbk. : alk. paper)
1. Fishing—New York (State) Guidebooks. 2. New York (State) Guidebooks. I. Luftglass, Manny, 1935– . II. Title.
SH529.B45 1999
799.1'09747—dc21 98-25679
CIP

British Cataloging-in-Publication data for this book is available from the British Library

Manufactured in the United States of America

This book is dedicated to thousands of people who keep the fishing in New York State among the best in the nation. These begin most prominently with the men and women of the New York State Department of Environmental Conservation (DEC), who creatively manage resources and an incredible array of programs ranging from stocking and fishery management to building artificial reefs. These also include others in the Department of Environmental Protection (DEP) who work to maintain the environmental integrity of our rivers, lakes, and bays. And certainly they include members of literally hundreds of conservation organizations and fishing clubs who make their own key contributions to the quality of fishing in the Empire State.

This book also is dedicated to those who helped us in creating and verifying its contents, most especially Patrick J. Festa, supervisor, Inland Fisheries, and Bob Brandt, supervising aquatic biologist, both for the DEC's Bureau of Fisheries, who combine a real love of fishing with equal zeal for making the fishing better for the rest of us.

In addition, we acknowledge with gratitude the contributions of a number of fishing writers (especially the writers and editors of the excellent "Good Fishing" series of books), plus editors, party-boat skippers, and other experts who provided invaluable information and insights.

Finally, but most profoundly, this book is dedicated to our departed fathers, Sam Bern and Harry Luftglass, who introduced us to fishing the right way: with expertise, with patience, and with affection.

Contents

Illustrations

Foreword

It is May, a tough time for New York anglers. What to do with a three-day weekend? Twenty- to 30-pound-plus striped bass are still gobbling live or cut herring in the lower Hudson River. American shad in the 2- to 8-pound range are thrilling light-line anglers and fly-casters with spectacular leaps at the Troy Dam and in the Delaware. Surf casters are frantically catching and releasing schooly stripers and some early blues at Montauk, and the fluke have shown in good numbers in the sound. Jiggers and drifters are taking big fat walleye in the St. Lawrence and Susquehanna Rivers, and in Conesus and Saratoga Lakes. Flat-line trollers are catching surface-walking Atlantic salmon in Lake George and Cayuga Lake and chunky 3- to 6-pound browns within casting distance along the entire Lake Ontario shoreline. Out "west," bass lovers are enjoying another super spring catching and releasing 2- to 5-pound smallmouth bass, one after another, in the special Lake Erie early bass season. More relaxing but no less enjoyable, local lakes and ponds are dotted with boaters and waders fly-rodding prespawning concentrations of big bluegills or ultralight jigging for a tasty meal of fat crappie or yellow perch.

Decisions, decisions. Oh, and not to forget that the ice is out in the Adirondacks and hundreds of wilderness ponds are teeming with eager and beautiful native brook trout. Plus the word is in that those big wild browns in our storied Catskill streams are already taking dry flies. Meanwhile lots of ferocious northern pike are exploding out of weedbeds to engulf spoons and spinners in Lake Champlain, while big and toothy tiger muskellunge are cruising for food in the Mohawk River, and Otisco and Greenwood Lakes. The truly mind-boggling options go on and on. Fortunately the regular bass and pure-strain muskellunge seasons don't open until mid-June.

When describing New York's amazing list of sportfishing opportunities to people living outside the Northeastern United States, one encounters surprise or skepticism, because to the uninitiated the state is often considered just a backdrop for the urban, hustling and bustling city. However, to those who have fished here, the above litany is pretty much preaching to the choir. With some four million acres of fresh water and 1,800 miles of Long Island saltwater shorelines, New York offers, I'm quite sure, more diversity in types of quality recreational fishing than any other state in the Union.

To perpetuate and enhance these resources, the Bureau of Fisheries staffs nine management offices, two research stations, and twelve hatcheries across the state with 200 dedicated people. The Bureau of Marine Resources, with offices and research units on Long Island and in the lower Hudson Valley, maintains a staff of 70 professionals to monitor and manage our marine fisheries. Additionally, 60 people in our Bureau of Habitat work diligently to protect the wetlands and aquatic habitats that support these fish populations. New York anglers are provided with over 1,200 miles of public stream fishing rights with 250 parking areas and nearly 300 boat-access sites for our "fishy" waters. Nearly one million pounds of trout and salmon are stocked each year to augment wild fish populations and create special sportfishing opportunities, including the chance to catch 20- and 30-pound chinook salmon and steelhead of 10 to 15 pounds in Lake Ontario and Lake Erie and in many of their tributary streams.

The New York State Department of Environmental Conservation provides much of the scientific and facilities infrastructure to maintain this fishing paradise, but the million-plus anglers who ply New York's waters provide the spirit and the involvement that keep it alive and special. Individually and in hundreds of angling clubs and organizations, they work to keep the financial and legislative support necessary to protect and manage these resources. They volunteer time to many conservation endeavors, from angler diary programs to stream improvement projects to kids' fishing derbies, to cite but several. Individually they practice ethical fishing tenets in respecting private property, not littering, complying with management regulations, and in other common courtesies that help maintain access to our streams, lakes, and ocean shores and assure everyone enjoys their days on the water. Catch-and-release habits are becoming increasingly the norm in many fisheries, and this conservation practice, as much as anything, is responsible for the high proportion of large and trophy-size fish found in so many of New York's sportfish populations.

Bottom line: Just about any type of recreational fishing endeavor you want to pursue—from bluegill to bluefin—is waiting for you in New York waters. There are a lot of friendly and fish-minded people waiting to welcome you to the fold.

Gerald R. Barnhart,
Director, Division of Fish, Wildlife,
and Marine Resources,
New York State

Authors' Preface

Ron Bern grew up fishing the pristine lakes and rivers of Southwestern South Carolina and moved first to Manhattan and then to Long Island in the late 1950s. Manny Luftglass was born and raised in Brooklyn and grew up fishing the fresh and salt waters of the Empire State. We met when we moved our families to the same neighborhood in central New Jersey and soon began a fishing partnership that has lasted more than a third of a century.

People from other parts of the country tend to think of New York State in terms of concrete, skyscrapers, and crowded highways. We know how incomplete this view is. We have experienced the sheer beauty and bounty that is New York outdoors, and we are both better men for it. This book is written to share what we have learned about fishing some of the 4,000 lakes and ponds, the 70,000 miles of rivers and streams, and the great coastal fisheries of New York.

We have examined only 100 spots here and we acknowledge that this merely scratches the surface. However, our objective was to make this a representative sample, including large rivers and small streams, massive lakes and small ponds, surfcasting spots and saltwater bays, artificial reefs and bluewater canyons. We think we got it about right. And as a little bonus, we've added our own private tips that can help you catch more fish.

We generally fish together, just the two of us. But this time we would like you to come along with us, beginning with the storied trout streams of New York and continuing through the best of marine fisheries.

Gone Fishin'

Key to Freshwater Fishing Spots

1 Ausable River, West Branch
2 West Canada Creek
3 East Koy and Wiscoy Creeks
4 Kinderhook Creek
5 Esopus Creek
6 Delaware River, West Branch
7 Willowemoc Creek
8 The Beaverkill
9 Callicoon Creek
10 Lake Champlain
11 Lake Ontario, Western Basin
12 Lake Ontario, Eastern Basin
13 Lake Erie, Eastern Basin
14 Conesus and Honeoye Lakes
15 Keuka Lake
16 Seneca Lake
17 Cayuga Lake
18 Owasco Lake
19 Skaneateles Lake
20 St. Lawrence River/Thousand Islands Region
21 Niagara River, Upper
22 Niagara River, Lower
23 Hudson River, Upper
24 Susquehanna River
25 Delaware River, Main Stem
26 Hudson River, Lower
27 Chateaugay and Salmon Rivers
28 Oswegatchie River
29 Salmon River
30 Mohawk River, Lower
31 Genesee River
32 Cattaraugus Creek
33 Delaware River, East Branch
34 Neversink River, Main Stem
35 Croton River, West Branch
36 Connetquot River
37 Peconic River
38 Cannonsville Reservoir
39 Pepacton Reservoir
40 Ashokan Reservoir
41 Rondout Reservoir
42 Swinging Bridge Reservoir
43 Titicus and Muscoot Reservoirs
44 Croton Reservoir
45 Kensico Reservoir
46 Black Lake
47 St. Regis Canoe Area
48 Saranac Lakes
49 Tupper Lake
50 Schroon Lake
51 Lake George
52 Oneida Lake
53 Great Sacandaga Lake
54 Saratoga Lake
55 Chautauqua Lake
56 Greenwood Lake
57 Clove Lakes
58 Lake Ronkonkoma
59 Fort Pond

New York State

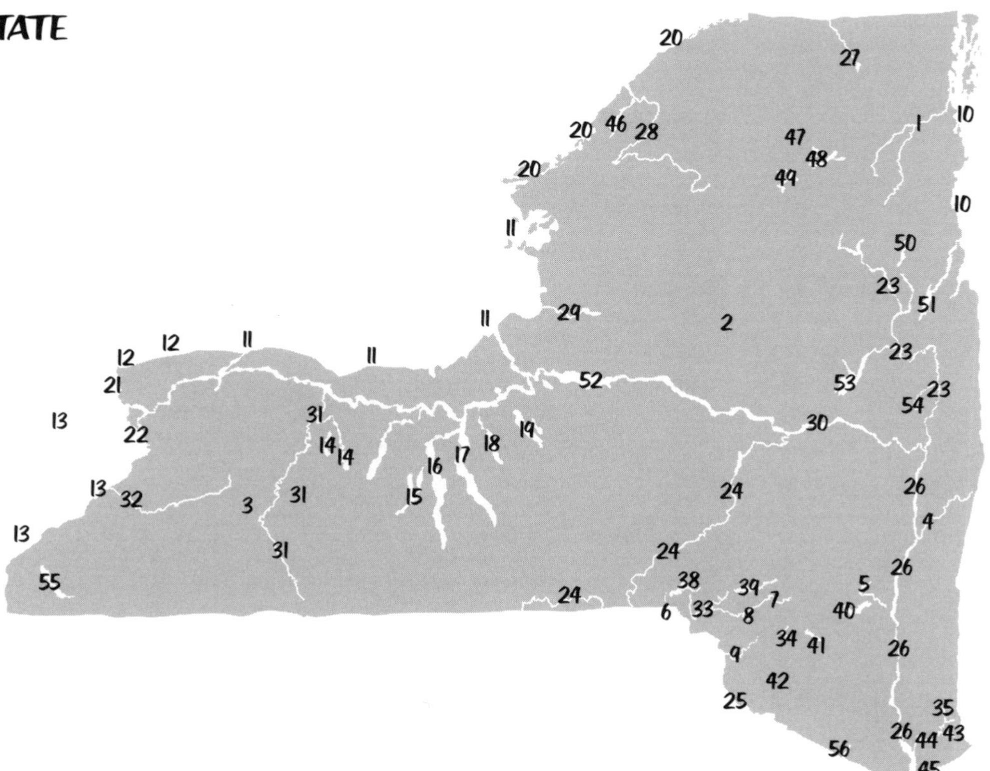

Key to Saltwater Fishing Spots

60 Mud Buoy
61 B. A. Buoy
62 Seventeen Fathoms
63 Cholera Banks
64 Hudson Canyon
65 Block Canyon
66 Block Island
67 Coxes Ledge
68 Rockaway Reef
69 *WAL-505*
70 *Arundo* Wreck
71 Town of Hempstead Reef
72 *Resor* Wreck
73 Fire Island Reef
74 Airplane Wreck
75 Texas Tower
76 Great South Bay Reefs
77 *Hylton Castle*
78 Tug and Tow Wreck
79 *Virginia* Wreck
80 *Coimbra* Wreck
81 Raritan Bay
82 Gateway National Recreation Area
83 Hoffman Island Sand Bar
84 Norton's Point
85 Marine Parkway Bridge Area
86 Breezy Point Jetty
87 City Island
88 Western Long Island Sound
89 East Reynolds Channel
90 Great South Bay
91 Jones Inlet, Outside
92 Fire Island Surf
93 Moriches Bay
94 Shinnecock Bay
95 Shinnecock Canal
96 Shinnecock Inlet
97 Shinnecock Inlet, Outside
98 Peconic Bay
99 Orient Point
100 Montauk Rips

Coastal New York

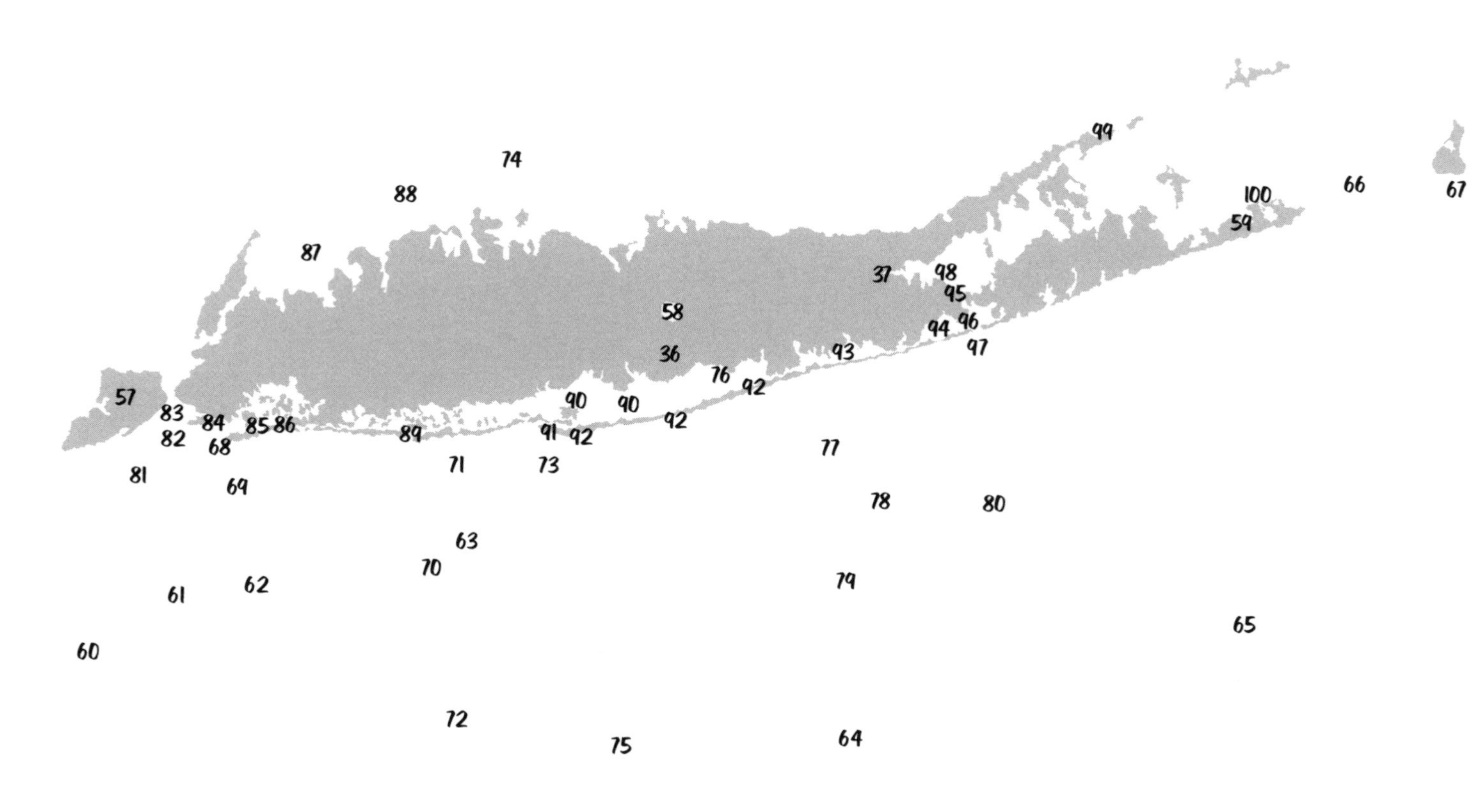

Freshwater Fishing

Fabled Trout Streams

The great trout anglers of the past 150 years have fished New York's superb trout streams with something approaching awe. The pools on these streams—for example, the famous "Junction Pool" on the Beaverkill and "Van Aken's Pool" on the Willowemoc—are known today by the name such angling greats as Theodore Gordon gave them at the turn of the century. The streams themselves—the Beaverkill, Willowemoc Creek, Esopus Creek, Neversink River, the West Branch of the Ausable River, the West Branch of the Delaware River among them—are still fished with a kind of reverence that we believe they deserve.

1 Ausable River, West Branch

Directions: Take Whiteface Mountain Memorial Highway, which crosses the river at Wilmington and intersects with Route 86. Follow Route 86 along much of the upper West Branch. Route 9N tracks the lower section of the West Branch, from Au Sable Forks downstream to the junction with the East Branch.

The main stem of the Ausable River and its East Branch both are fine fisheries in their own rights, with abundant populations of bass, pike, walleye, and trout. However, it is the West Branch that anglers dream about when they think of trout. Called by some the number-one trout stream in the eastern United States, the West Branch is a moveable feast in the fullest sense of the word, providing more than 25 miles of prime trout water. Trout anglers—many armed with fly rods—trek from all corners of the nation to pit their knowledge and skills against the brown, rainbow, and speckled trout that flash through the gin-clear waters of the river.

The West Branch is born in the marriage of several high mountain brooks just south of the village of Lake Placid. In its first 5 miles, it quietly gathers volume from small feeder springs until it reaches the gorges at Wilmington Notch. At this point, the river is perhaps its most beautiful, tumbling over a waterfall that is more than 100 feet high. At the bottom of High Falls, the river has gouged out a deep pool surrounded by rocky ledges. It is from these ledges that anglers work the pool and often take lunker trout.

Several miles farther downstream, the river cascades over Flume Falls, beyond which the river flows quietly into a dammed-up section called Lake Everest. The several miles above this small lake and the lake itself are first-rate baitfishing areas for big browns. The section below the dam features tempting pockets and pools, which are alive with trout, and which fly anglers ply with fine success. Another section of the river is dammed at Au Sable Forks, and this quiet water also holds trophy fish.

Along its route, the West Branch flows through country so beautiful as to make the trip worthwhile, whether or not the angler finds fish. For fly anglers seeking to "match the hatch," the progression of fly hatches proceeds from Hendrickson in May through Caddis, Grey Fox, March Brown, and Green Drake in June, Light Cahill and Tricos in July, and Isonychia in late August through mid-October.

The West Branch is largely subject to New York State trout regulations, including numbers, size limits, and closed seasons. However, as of this writing, from the Route 86 bridge downstream to the Wilmington Dam, the season is year round with no size limits. From Monument Falls downstream for 2.2 miles, there is a "no-kill" section, which may be fished with artificial lures only.

Tip: Although the romantic version of fishing the West Branch involves the perfectly presented dry fly, spin casting artificials is another deadly way to put fine fish in your creel. Our favorite is a small Mepps spinner in gold or silver; others prefer a small Rapala or a Panther Martin. When casting spinners, cast quartring upstream. Keep the rod tip low, point it at the lure, and retrieve it fast enough to keep the spinner from snagging bottom. In deep holes, however, let the spinner sink near bottom, then twitch the rod tip to get the blade rotating.

The West Branch of the Ausable combines the best that nature and a highly effective Department of Environmental Conservation (DEC) can provide to assure fine trout fishing well into the future. Running through a cold section of the Adirondacks, the river is very clean, well-oxygenated, cool even in warmer weather, and full of the kinds of insects and small baitfish that cause trout to thrive. Moreover, the river is regularly stocked with thousands of brown trout in the 8.5- to 9-inch range at various points along the river, including Jay, North Elba, and Wilmington. This augments the river's already healthy stocks of trout.

It should be added that the main branch of the Ausable is heavily stocked with landlocked salmon and steelhead, and that the East Branch receives regular stockings of thousands of rainbow and brown trout. As noted in the beginning of this chapter, both the main stem and the East Branch are themselves fine destinations for sport anglers. However, it is the West Branch that is the classic of the three—notable for not only its exceptional trout fishing, but also its exceptional beauty.

Whereas most anglers fish the West Branch from shore or wading, there is a boat launch just off Route 86 on Lake Everest, just above the intersection with Route 431.

2 West Canada Creek

Directions: N.Y. Route 28 follows the streambed of the West Canada through most of its journey through Herkimer and Oneida Counties.

When well-traveled trout anglers are asked to rate the great streams of New York, West Canada Creek is always near the top of their lists. This ranking is based not on great trout populations alone, although trout are found in virtually every stretch of water from beginning to end. It is not based on sheer beauty alone, although this little river is certainly one of the most beautiful in the state. Nor is it based on accessibility alone, although the West Canada can be fished along virtually its entire length, and in addition, a good paved road parallels most of the creek's impressive length. But all of these factors—taken together with the deep pools and long stretches of riffles and tempting pocket waters that dot the creek—demand the respect that West Canada Creek has rightfully earned.

The creek begins as the outflow of lakes, the larger being West Canada Lake in eastern Oneida County. The South Branch, which joins the main river near Nobleboro, is surrounded by state lands and is itself a fine little trout stream. Also largely surrounded by lands owned by the state, the main branch provides classic trout waters for much of its length.

West Canada Creek is a year-round trout fishery. The largest fish are characteristically taken in spring; however, the creek's many deep spring holes provide good trout fishing in summer when other, shallower creeks have quit producing.

Although the creek also holds brook trout, rainbows, and surprisingly large bass, it is brown trout that have classically attracted anglers to the West Canada. Browns favor cool, well-oxygenated water with convenient food resources; consequently the inlets of small feeder streams along the main branch are ideal spots to fish. The riffle sections are often very productive, especially in spring and early fall.

When fishing the West Canada, one must be prepared for the slashing strike of a big brown, since the creek harbors many very large fish. Remember that a big trout's first instinct is to dash for cover, and in the process, it may quickly wrap the line around a rock or other snag to break free.

A superb trophy-trout section is found below Trenton Falls, extending about 2.5 miles downstream to the mouth of Cincinnati Creek. Only artificial lures are allowed in this stretch and creel limits at this writing are a maximum of three fish 12 inches or longer.

Tip: When wading the trophy section of the creek, be vigilant for sudden rising water levels. There are no set schedules for water releases from the power-generating plant at the dam at Trenton Falls. However, sudden releases of water at the dam can threaten the safety of anglers concentrating on the next cast.

Of the destinations along the West Canada, anglers are especially attracted to conveniently reached stretches in the vicinity of Poland, Middleville, and Hinckley. Fine fishing is found near parking areas below Poland and in fishing areas between Newport and Middleville that are situated just off Route 28.

The West Canada supports a healthy native population of browns, liberally augmented by often sizeable trout stocked each year by the DEC. In 1997, for example, the DEC stocked 32,280 browns in the creek in Herkimer County, ranging in size from 8 to 14 inches. In Oneida County, another 7,520 browns were stocked, ranging in size from 8 to 14 inches, with a few trophy fish as large as 17.5 inches. Several hundred brook trout also went to these waters.

Most anglers fish either wading or from the banks of this rich and historic trout venue. However, the occasional canoeist works the deeper pools and larger channels of the creek.

3 East Koy and Wiscoy Creeks

Directions: Take NY Route 39 in Wyoming County to Lamont; then take Lamont Road north or south to access the East Koy. Wiscoy Creek is reached via Route 39 near the town of Pike Five Corners.

East Koy Creek flows cold and clean between brushy tree-lined banks and rolling meadows toward its juncture with the Genesee River. A small stream, the East Koy is often considered one of the most underrated trout streams in Western New York State, perhaps rivaled only by its sister stream the Wiscoy.

One of the most noted stretches of the East Koy is found in the vicinity of Goldenrod Campground upstream from Gainesville. In this section, the creek flows 2 to 3 feet deep, largely over gravel beds, and brown trout abound. It is in this general area that the DEC concentrates its stocking activities. In a recent year, 13,500 brownies ranging from 8 to 14.5 inches were added to the river's stocked and stream-bred trout populations in four separate stockings. Because of the excellent quality of the water, stream reproduction also is an important element in the stream's trout profile.

While many spin anglers fish the East Koy, this little stream is essentially a fly angler's paradise, with plenty of riffles, pools, and pocket water to work and extraordinary hatches to match, especially a phenomenal caddis hatch each year.

Tip: During high-water early season stream fishing, it is generally necessary to get lures (or bait) deep, as close to bottom as possible. As water warms up and the stream level drops, trout will hit at all levels, with flies and spinners becoming more effective in enticing trout to strike.

The Wiscoy springs up in southern Wyoming County, beginning as two separate branches, which meet at the town of Bliss. It then flows southeasterly toward its juncture with the East Coy and ultimately the Genesee River. Along the way, it is enriched with cold, oxygenated water from several tributaries, the most notable being Trout Brook.

The Wiscoy presents classic fly-casting waters virtually from its headwaters to its juncture with the river. At its upper end, an open stream flows through gently undulating fields, providing adequate flows, occasional deeper pools, and very pretty varied water. At the town of Bliss, a 1-mile no-kill stretch is a favorite among local "pros," who agree with noted fishing authority Lee Wulff that "a game fish is too valuable to be caught only once."

Naturally reproducing brown trout in the Wiscoy and its principal tributary, Trout Brook, make stocking unnecessary. The water remains cold right through the summer along its upper reaches, providing respectable trout opportunities when other streams have long since quit. In fact, hatches of cahills, slate drakes, and tricos keep the Wiscoy productive for anglers right through the month of July. Not surprisingly, there is more than a little fishing pressure on this beautiful, productive and approachable stream.

It should be added that the Genesee River, the terminus of the East Koy, is itself a wonderful trout fishery. Whispering like a gift out of Pennsylvania, this beautiful river flows north through western New York on its way to Rochester and Lake Ontario. The Genesee is liberally stocked each year with brown trout and rainbows, from the Pennsylvania state line downstream to Belmont Dam. Its upper stretches are especially prized by trout anglers, although the lower section benefits dramatically from flows of several excellent trout streams, including the merged East Coy and Wiscoy Creeks. A 2.5-mile catch-and-release stretch at Shongo just north of the New York/Pennsylvania line is one of the most attractive and productive trout stretches on the river.

Because of its cold, clean waters, the Genesee also offers a fine smallmouth bass fishery in the area of Wellsville, which has the additional benefit of being the river's heaviest traditional trout stocking point.

Both the East Koy and the Wiscoy are readily accessible. Campers in the Goldenrod Campground on Shearing Road upstream from the town of Gainesville have automatic access and the owners of the campground are often quite gracious in granting permission to fish there. The East Koy also is accessible from various points off Lamont Road off N.Y. Route 39. The Wiscoy is readily accessible off N.Y. Route 39, although a bit of care is required in seeking permission to fish certain posted stretches. N.Y. Route 19 tracks closely along the Genesee from the Pennsylvania line all the way up to the junction with the East Coy. Take Graves Road near Shongo to the river's longest catch-and-release stretch.

4 Kinderhook Creek

Directions: Take U.S. Route 90 across Kinderhook Creek near the Columbia County/Rensselaer County Line. Follow Route 66 across the creek and down its shoreline to Chatham, where the stream doubles back upon itself. Take Route 9 along the lower stream to its junction with the Hudson.

Kinderhook Creek is a fine trout stream in the Capital District, located so close to Albany (some 20 miles as the crow flies) that one imagines dedicated anglers among legislators and businesspeople looking out of office windows with increasing restlessness as spring arrives. This excellent stream, formed by a junction of brooks in Rensselaer County, winds a sinuous path through Rensselaer and Columbia Counties before joining with the Hudson River. Along its 45-mile path, it provides a variety of angling opportunities, but when one considers Kinderhook Creek seriously, it is brown trout that rivets the attention.

The primary trout waters are in the upper half of the creek, stretching downstream from the junction of East and West Brooks at Stephentown for about 20 miles to the area below Malden Bridge. (Additional trout possibilities in several miles of the West Brook should not be overlooked.)

The prime trout waters of the Kinderhook provide every possible condition conducive to trout, from rock-strewn rapids to deep, shaded pools; from fast riffles and pocket water to classic rock-and-gravel-bottom runs. Conditions are ideal for fly and spin casting. The most productive lures here are very small spinners and a variety of flies that match the hatch, prominently to include Hendricksons, caddis, and blue-winged olives.

Tip: Although fly casting admittedly appeals to our aesthetic senses, it is our experience that light spin casting with bait in the slower, deeper pools attracts most of the largest trout. Worms are most effective, worked through pools and bounced along the rocky bottom of riffles to emulate natural foods in the current. Tiny minnows also are effective in some of the deeper pools. The emphasis in this environment of sophisticated trout is "natural presentation."

A number of state easements along the river allow public angling access, including stretches of the river at Adams Crossing and Gould Road. Within an easy stroll of these points are deep pools that produce trout on a regular basis. Not surprisingly, they attract a good deal of angling pressure; nonetheless they produce fish, especially when fished during the quieter hours at sunup and after dark.

The creek is crossed by Route 66 in several places, and the ready access provided there also attracts anglers. Upstream, especially where the Kinderhook joins Black River, there are quiet pools where one can generally fish in solitude.

The DEC works assiduously to assure the continued quality of trout fishing in this great creek. In a typical stocking year, 6,850 brown trout—ranging from 8 to 14.5 inches—were stocked at Chatham and New Lebanon in Columbia County. More than 8,000 additional browns in the same size ranges were stocked at Nassau and Stephentown in Rensselaer County.

5 Esopus Creek

Directions: Take the New York State Thruway to Exit 19. Follow N.Y. Route 28 West. Route 28 parallels most of the 11 miles of the "Big Esopus," the best and most fishable section of the creek.

Esopus Creek is one of the premier natural rainbow trout hatcheries in the Eastern United States and certainly one of New York State's most attractive trout streams. Situated in the Catskill Forest Preserve, the source of the Esopus is Winisook Lake in Ulster County. In its first miles, the "small" Esopus is of modest proportions. Several miles of the "small" Esopus immediately below Winisook Lake are on NYS Forest Preserve and are therefore accessible by means of a hike. The Esopus then becomes heavily posted down to Big Indian.

Several spring-fed tributaries, including Warner Creek and Birch Creek, swell the creek's volume (and provide their own fine trout-fishing opportunities in spring and early summer). At "the Portal," the 18-mile long Shandaken Tunnel empties cold Shoharie Reservoir waters into the creek. Here the "large" Esopus begins. From this point, the creek flows through 11 miles of beautiful and easily accessible riffles, pools, and pocket waters to its appointment with the fish-rich Ashokan Reservoir.

Populations of wild, naturally reproducing rainbow trout are regularly augmented with state stocking of browns, such that DEC surveys have discovered that trout populations outnumber coarse fish in the Esopus. This is most desirable.

We fished the Esopus together on a fine October morning in 1998, a short walk though the woods from The Lodge at Catskill Corners, a lux-

Figure 1 ■ Trout angler Hal Anderson shared bait and extensive knowledge about fishing Esopus Creek with Ron and Manny. *(Photo: Ron Bern.)*

urious new hotel we chanced to find on Route 28 below Phoenicia in our wanderings in the Catskill Forest. The stretch we chose to fish—a long pool at the foot of a series of rocky ledges and riffles—was a short walk through the woods directly behind the lodge. It was a spot we both knew held trout. However, an hour's work with a wide variety of artificials produced not the first rise.

At the moment we were considering a move, a silver-haired gentleman named Hal Anderson appeared with his regular fishing partner, a handsome springer spaniel. Mr. Anderson, who lives in nearby Shokan and who trout fishes the Esopus on a regular basis, regarded our offerings of artificials for a minute or two before advising that "today these trout want fresh salmon eggs." His advice took on a certain urgency when a fine rainbow took his first offering. As he gently unhooked his fish, he smiled and said he had more than enough bait and invited us to share it. When we availed ourselves of his generosity, we experienced a strike on virtually every cast, with RB catching and releasing four sassy but regrettably small native "bows" on four consecutive casts. Our new friend explained that the run of larger fish was yet to come upstream from the Ashokan Reservoir.

Tip: Of all tips in this volume, this one is the most obvious. Use a #8 or #10 hook, a light weight, and a small glob of fresh salmon eggs. When baiting your hook, wrap the eggs around and around the hook at least a dozen times so that the light connective tissue between the eggs will keep the bait on your hook.

An hour later we were drinking coffee with another transplanted trout fisherman, an authority on the Esopus named Bill Tallon. Bill catches good-sized rainbows in the spring as they run back into the creek from the Ashokan to spawn. A bonus is the occasional autumn-run brown trout, which also may reach impressive size and like the rainbows, fights well in the cold water.

Bill also fishes salmon egg clusters in the creek. However, his experience is that brown trout egg clusters work far better. Of course, countless trout are taken on artificials, ranging from hatch-matching flies to the small silver Mepps spinners our friend Hal Anderson likes to cast. Mealworms, garden hackle, and standard (jarred) individual salmon eggs take their share of trout as well.

Esopus Creek is generally regarded as a small fish stream, with the average size of rainbows and browns clustering between 9 and 10 inches. That's because most of the "bows" migrate to the Ashokan Reservoir when they reach between one and two years of age. However, these trout begin to migrate back up the Esopus when they reach about 16 inches in size the following year. The average fish making the journey up the creek to spawn in springtime is between 16 and 22 inches in length—usually when the weather is still quite cold. In mid-autumn, spawning browns ascend the creek in numbers, mixed in with a few large rainbows. Browns of 18 inches or more are not unheard of, especially during a 30-day period from mid-October to mid-November. In all seasons, the sheer abundance of trout plus the beauty and accessibility of the Esopus rank it with the best trout streams in the country.

As if this weren't enough, there is a springtime hot spot for spawning walleyes near the creek's juncture with the Ashokan. Called "Chimney Hole," this large and quite beautiful pool produced a state record in 1923 that stood for more than three decades: a spectacular 19-pound 4-ounce brown trout. Today Chimney Hole holds what may be lineal descendants of that great fish, along with rainbows and walleyes as large as 14 pounds and sizable smallmouth and largemouth bass. As Shakespeare might have said, "'Tis a consummation devoutly to be wished."

6 Delaware River, West Branch

Directions: Take N.Y. Route 17 (Quickway) from Hancock to Deposit.

Our friend Howard Brant, a hunting and fishing editor for the *Newark Star-Ledger*, once called the West Branch the finest trout water in the Catskill region. He noted that veteran anglers consider the West Branch, especially from Deposit to Hancock, the "crown jewel of the Delaware River Valley's wild trout fishery, akin to the best wild trout waters in the country." He certainly knows whereof he speaks.

The upper reaches of the West Branch support fine self-sustaining populations of brook and brown trout as it flows through miles of gently rolling farmland, from its headwaters near the slopes of Utsayantha Mountain south to a point between the villages of Walton and Deposit, where the river is impounded to form the Cannonsville Reservoir. In addition, the DEC stocks trout in the branch each year to bolster native populations. For example, more than 18,750 browns were stocked in the branch in 1997. The smallest of these were 8.5 inches; the largest 14 inches.

The construction of the reservoir affected the nature and quality of fishing in the West Branch quite dramatically. Previously, trout were found almost exclusively above Deposit. However, cold-water releases from the bottom of Cannonsville now create a trout fishery virtually the entire length of the branch. (As noted elsewhere, the reservoir is itself a great trout location, with anglers catching brown trout there as large as 20 pounds.)

The river below Cannonsville Reservoir is shaped roughly like a fishhook, with the barb set in the lower end of the impoundment and the eye at the beginning of the main stem in Hancock. Water conditions vary significantly with the rainfall and associated drawdowns from the reservoir. In early summer, depending upon weather conditions, the river may be low and weedy. However, heavy summer releases of water through the warmest months keep the river cold and generally full.

Tip: Rising water may be a good indication that trout are starting to feed. Rains wash insects and other foods off streambeds and overhanging trees, and stronger currents dislodge insect larvae and other morsels from bottom. Trout begin feeding when swirling water carries food past their lies.

The occasional smallmouth bass may be found in the West Branch. However, this fine river, in the reaches above the Canonnsville Dam and especially in the cold water below the dam, is trout country—not only because of the extraordinary numbers of browns, rainbows, and brookies supported by the river, but also owing to the size of the fish, most especially the browns. West Branch brown trout average 13 to 15 inches, with fish of 20 inches or more not unusual. Occasionally, fly anglers find 5- to 7-pound browns tearing line off their reels.

Much of the land along both sides of the West Branch is in private hands; however, landowners often are quite gracious in granting fishing access to those who ask permission. Some public-access points do exist, including a popular stretch on both sides of the river from Deposit downstream almost to Hale Eddy. This stretch is still owned by New York State. Another popular access point is right at the Hale Eddy bridge, complete with good parking and excellent pools and riffles to test the angler's skill.

Below Hale Eddy the Commonwealth of Pennsylvania has acquired several miles of riverbank, which can be fished with either a New York or Pennsylvania license. In addition, Pennsylvania maintains a nice access point at Balls Eddy, including a launch ramp for canoes and plenty of parking.

7 Willowemoc Creek

Directions: The Willowemoc flows generally along Old Route 17, which provides a number of parking areas for anglers.

When dedicated fly-casters speak of the great eastern trout streams, a few names are mentioned almost with reverence. One of these is Willowemoc Creek, where such trout-angling greats as Theodore Gordon and Lee Wulff cast their hand-tied flies to wily native trout.

Like its sister river, the equally famous Beaverkill, the Willowemoc is generally perceived as two sections. The upper section, sometimes called the "Little Willowemoc," is fed by clean, cold mountain streams—principally Fir Brook and Butternut Brook. These and other brooks together form the main branch of the creek, and at the point where Fir and Butternut enter, anglers find wild brook trout in surprising numbers (as well as in the deeper holes of the main feeder streams). There is considerable state land here, and more angler access than one finds on the upper Beaverkill.

There are stretches of private water below this point; however, from just below the hamlet of Willowemoc to a point several miles downstream, the state has secured easements for anglers. Private and public stretches alternate, with more easements providing access from below Mongaup Creek for a number of miles to the Quickway Bridge. The river gains in size and strength here, and the trout get larger, too—especially the browns, which run to 18 inches and more.

The "Big Willowemoc" is a gorgeous, classic fly-fishing stream, beginning at Livingston Manor and flowing to the famous "Junction Pool," where the Willow joins the Beaverkill. Much of this stretch of the river is accessible via state easements, hatches are very good, and the water is large enough for casting without being outsized like the West Branch of the Ausable or even the Beaverkill. There is a special "no-kill" section of the Big Willowemoc where only artificial lures may be used, and there is no closed season. This 2.4-mile stretch beginning at Bascom Brook yields some of the largest trout to be found anywhere in the creek, in spring and fall, of course, but also on mild winter days when most other trout fishing is prohibited. Among the famous pools in this stretch are Wegman's, Hazel Bridge, Trestle, Power Enclosure, and Van Aken's Pools. (The famed Catskill Fly Fishing Center is located right on Van Aken's.)

Tip: Remember that wariness is a trout's number-one survival mechanism. Sudden movements, noise, vibration, or even shadows will send these fish darting for cover. Sneak up to the streambed, keep movement to a minimum, take extra care in presenting your fly or lure. If you accidentally spook trout, move on to another spot, because the one you frightened won't be feeding for a while.

The creek is located partially in the 300,000-acre Catskill Forest Preserve, which is managed under the "Forever Wild" clause of the state constitution. This makes the creek and other waters in the preserve easier to protect.

The DEC manages the Willowemoc with special care, stocking browns both in quantity and in quality. In a recent stocking year, 22,160 browns ranging from 8.5 to 14 inches were stocked in the upper portion of the creek at Neversink, and 13,820 browns ranging from 7 to 14 inches were put in the lower ("Big Willowemoc") creek at Rockland.

The large trout in the Willowemoc and the romantic mystique that hangs over its history call to mind some lines written by fly angler Frederick White:

> If, occasionally, in fast and dangerous water or on some tumbling, brush-hung brook, the angler strikes and plays and nets a really heavy trout, he has proved his nerve and skill and sureness of touch and may well regard his success as the result of patient and hard-won experience. But whatever his fortune, it is safe to assert that he will continue to seek the unknown and the well-nigh unattainable. For as long as imagination tempts and hope persists there will remain that undiscovered star of the angler's firmament, that biggest fish of all—the one that gets away.

8 The Beaverkill

Directions: Take Route 206 into Roscoe. Follow Route 17 from Roscoe downstream for the entire length of the lower Beaverkill.

The Beaverkill has been called the standard by which all other trout streams are judged. Steeped in tradition and lore, this 42-mile stream has been the favorite of America's most gifted fly anglers for 150 years and still acts today as a magnet to men and women everywhere who fly-fish for trout. Legendary trout angler Corey Ford once wrote, "A trout stream is more than the fish in it. A great trout stream like the Beaverkill is a legend, a fly book filled with memories, a part of the lives of all the devoted anglers, living or dead, who ever held a taut line in its current." Certainly he expressed the sentiments of countless thousands who fish and have fished this extraordinary stream.

The wellspring of the Beaverkill is a narrow, rocky ravine between two mountains deep in the Catskill forest. Water as sweet as springtime seeps into an ancient beaver meadow, where small native brook trout flit through the shadows. As it continues westward for another 12 miles, the stream is enriched by dozens of springs and feeder streams. At Touchmenot Mountain, the stream turns southwesterly and flows another 14 miles to its junction with Willowemoc Creek at the renowned Junction Pool near Roscoe. The waters of the Willowemoc, considered by many the equal of the Beaverkill, double the Beaverkill's size and mark the beginning of the lower Beaverkill. The pools are large and deep here, the riffle areas lengthy as the enlarged Lower Beaverkill flows toward its junction with the Delaware River's East Branch.

Unfortunately, virtually the whole of the upper Beaverkill is private waters, although access is available at the state's Covered Bridge Campground for a fee, and along a 2-mile stretch from the bridge at Route 206 downstream to Roscoe. (The "upper upper" Beaverkill is on NYS Forest Preserve land and accessible to the intrepid angler willing to hike in. The river here is a small stream populated mostly by brook trout.)

However, the lower Beaverkill—the so-called "big river," from Roscoe down to the East Branch of the Delaware—could hardly be more accessible, given DEC-acquired easements along virtually its entire length.

The most famous and productive pools, after Junction Pool, are Ferdon's Eddy, Barnhardt's Pool, Hendrickson's Pool, Horse Brook Run, Cairns Pool, Wagon Tracks, Schoolhouse Pool, Painters Bend, and Cooks Falls Pool, where water cascading over a large mass of ledge rock creates excellent trout habitat. Each of these pools has its own history and romance. Ferdon's Eddy, for example, was where Roy Steenrod tied a new fly to match an emerging hatch during a fishing trip in 1916. Day after day, he and his fishing partner caught scores of trout on this killer fly. As a tribute to his partner, A. E. Hendrickson, he named the fly after him. And so the classic Hendrickson fly came into being.

While the Beaverkill may summon thoughts of the past, it is very much a present-day phenomenon; still an extraordinary trout fishery for both wild and stocked browns. A significant number of some 27,000 browns stocked each year reach maturity. Relatively common are 14- to 18-inch fish, and those over 20 inches are very much in evidence.

Two special "no-kill" areas have been established on the Beaverkill; the first a stretch from a short distance below Roscoe downriver for 2.5 miles; the second about the same length from 1 mile upstream of the Iron Bridge at Horton to 1.6 miles downstream. Within these popular areas, artificial lures only are permitted, and all trout must be immediately returned to the water without unnecessary injury. Significantly, the proportions of wild trout are steadily increasing in these no-kill sections. Fully half of the trout taken here are stream-bred, whereas only a tenth of the fish taken throughout the river are wild.

Tip: There is good fishing below the no-kill stretches, both for fly- and bait-fishing. One especially productive bait pool is to be found from the Cooks Falls Bridge downstream for nearly a mile.

The legendary angler Lee Wulff was so taken with the Beaverkill during his first visit that he abandoned his other pursuits and moved to the area to establish his now-famous fly-fishing school on the upper part of the river. Of many graduates of that great school, the most memorable to us was a lovely third-year law student who related her days on the Beaverkill with her industrialist father, learning to get closer to fly-fishing, the stream, and each other.

As noted, there is some limited access to the upper river in its final several miles above Roscoe, notably from Cat Hollow Road downstream to Roscoe. (A state parking lot for anglers is located just off Route 206.) However, DEC-provided easements make access good from Roscoe to the Delaware River. (Note: These easements are across private lands, which means anglers must in some cases use certain marked parking or access sites, and restrict their activities to the stream and its banks.)

9 Callicoon Creek

Directions: Take Route 52 from the town of Jeffersonville along the main stem of the creek and most of its East Branch.

Callicoon Creek is a classic trout stream, widely known for long, perfect riffles and deep pools that harbor big brown trout. It is the destination of fly anglers from distant parts and with good cause. After all, it flows through beautiful Catskill countryside. It provides superb fishing in clean, pristine waters. And some of the most noted anglers, fly tiers, and angling philosophers of this century have dimpled its surface with their dry flies.

Thus Callicoon Creek's well-deserved image is almost exclusively a great trout stream. However, as we shall note a bit later, there are other images that some of us carry forth across long reaches of time; images no less fraught with pleasure or excitement than the handsomest brown trout rising to the most perfectly presented fly.

The headwaters of both branches of the creek rise up as tiny streams in the midst of Sullivan County. The north branch begins only a few miles away from the Beaverkill; the longer East Branch is closer to the Willowemoc and the Neversink Reservoir. They meet at the town of Callicoon to form the main stem, which flows into the Delaware River.

The Callicoon is a resource that is zealously protected and nurtured by the DEC. Each year, the DEC stocks yearling trout in both branches and the main stem of the creek. In a typical year, 1,000 browns were placed in the main stem, 3,900 browns in the East Branch, and another 3,000 in the North Branch—all in the 8.5- to 9-inch size.

Tip: If your trout is lightly hooked, use long-nosed pliers to quickly remove your hook and release the fish. If the fish is deeply hooked, simply cut the line without netting or handling the fish. Your hook will rust out in a short time and the fish will survive to fight again. Importantly, do this quickly, especially in warmer weather, since trout need to return to cooler, deeper water to recover.

To assure angler access, the DEC has secured perpetual fishing easements along the north branch of the creek, which allow fishing only in the stream and from the adjacent bank.

Of course, that is the situation today. But in writing about Callicoon Creek, we cannot help remembering another time, nearly a half century ago, when Manny spent long summer days fishing these waters with makeshift tackle rather than fine split bamboo fly rods, and in search of admittedly less-exalted fish than brown trout. But the differences in these images illustrate in satisfying ways how diverse and wonderful is the fishing experience for those of us who truly love it.

The slow-moving pools of the main stem were Manny's favorite fishing holes, reached after a pleasant walk down Maple Avenue in Jeffersonville from its junction with Route 52 in Callicoon. To the right was a bank, a movie theater, and a Rexall drugstore. A short distance straight down Maple was the old mill that received the flow from Lake Jeffersonville.

Chub (more than likely, true "fallfish") were the favorite quarry then, although we caught more than our share of rainbow trout, pickerel, smallmouth bass and largemouth bass, and extremely large sunfish and suckers. The basic equipment was a cane pole tied with black nylon thread. The baits were worms and bread dough, the latter molded into the shape and size of a nickel and sometimes sweetened with vanilla.

The pleasure of those days comes back in memories only magnified by time, and if anything, the fishing is even better now. So we have every intention of fishing Callicoon Creek together when our deadlines all are met. We know the old mill is shut down now, the A&P gone, the theater gone with it. But the essentials all remain unchanged: beautiful, slow pools, long stretches of riffles, neat undercut banks, and plenty of cool, clean water filled with fish to make memories for other days.

The Great Lakes

In their vastness and their richness, New York's Great Lakes—Lake Ontario, Lake Erie, and Lake Champlain, the "sixth Great Lake"—virtually defy brief description. The three lakes and their main tributary rivers have produced virtually every salmon and trout record in New York State. Spectacular northern pike, lake trout, walleye, and bass are commonplace catches in these waters. Some of the better-known fishing hot spots—Barcelona, Dunkirk, Sturgeon Point, and Cattaraugus Creek on Lake Erie, for example, and Wilson Harbor, Olcott, Oak Orchard, and Sodus Bay on Lake Ontario—each provides sufficient materials for a dozen books. In these brief pages, we can only hint at the resources of these Great Lakes.

10 Lake Champlain

Directions: Take Route 22 along the eastern shoreline of Lake Champlain in Essex County.

It is difficult to discuss Lake Champlain without superlatives. It is an immense body of water, with a surface area of 435 miles and 587 miles of shoreline. It is home to 83 species of fish, making it one of the most diverse freshwater fisheries in the country. It is beyond doubt one of the great "two-tiered" fisheries in the United States, with huge populations of such cold-water favorites as land-locked Atlantic salmon, lake trout, brown trout, and rainbows/steelheads, and equal populations of such warm-water fish as northern pike, pickerel, walleye, and bass. (Additional species, including sauger, whitefish, yellow perch, muskellunge, crappie, and even the occasional sturgeon ply the clean waters of Champlain.) The lake benefits from the clean, well-oxygenated waters (and stocks of fish) from some of the great fishing streams in the Northeast, of which the Ausable, Saranac, and Boquet Rivers are but three examples. Even the annual stocking programs are "larger than life," with as many as 450,000 yearling and fry Atlantic salmon, 125,000 lake trout, and 55,000 brown trout among the fish placed in the lake.

Classically referred to as the "sixth Great Lake," Champlain is 12 miles wide at its widest point and 400 feet deep at its deepest. Finally, the lake is a superb 12-month-a-year fishery. In the spring, when the ice goes out, the salmon come pouring into the mouths of the rivers, competing with big rainbow and brown trout for food washed into the warmer inlets and bays by runoff from melting snows. Lachute River and Putts Creek are among the most productive inlets at this time of year. Atlantics averaging 2 to 4 pounds are taken in good numbers (30-fish days are not unheard of) with fish running up to 8 pounds. At the same time, lakers averaging 4 to 5 pounds feed actively in deep water, often taking deep-trolled lures.

As May turns to June, salmon and trout move out from the river inlets to deeper water and strike most actively nearer the surface. Lakers spread out through all depths, including the occasional fish taken from shore in only a few feet of water.

At its southern end, the lake is narrow and shallow, with broad expanses of weedbeds and lily pads calculated to set the bass angler's heart aflutter. The king of these regions is the largemouth bass, which grows to fine proportions on the rich stocks of forage fish available. Rubber worms are perhaps the deadliest lure in the weedbeds. Of course, live minnows worked at the edges are productive, too. However, one is never quite sure who has come to call when a float is sucked under, since big populations of pickerel, walleye, yellow perch, crappie, and northern pike also inhabit these same waters.

A broad series of rocky points, riprap, and peninsulas hold wonderful populations of smallmouth bass, and spinnerbaits, crankbaits, and jigs are main elements in the angler's arsenal. Valcour, one of the key islands in the lake's broader expanse, is especially hot for bronzebacks as well as northern pike and walleye. The walleyes here are especially active by night, and big schools of yellow perch spice the live-bait action by day.

The mouth of the Great Chazy River near the lake's northern end, is another spring hot spot that acts as a magnet to anglers seeking muskellunge, as well as the whole spectrum of warm-water fish. Like the shoreline 100 miles to the south, this section of the lake features wide, weedy bays loaded with bass and pike.

Early fall marks the beginning of the land-locked salmon runs on tributary rivers such as the Ausable, Boquet, and Saranac, and the populations of big, healthy fish get better and better, thanks to the efforts of the Environmental Conservation departments of New York and Vermont to curtail the lamprey eel problem on Lake Champlain. Someone once said that catching a salmon on a fly is the most fun you can have standing up. And there is literally no place where the odds for success are better than the tributary rivers to Champlain.

Tip: Salmon in streams don't usually take the same lie that a trout might. As angler writer Fran Betters advises, "They will usually hold next to a boulder where the current is slower or at the tail end of a pool alongside an underwater rock. When lying in the head of a pool, they will most likely be along the edge of the current instead of out in the center."

In winter, ice fishing is exceedingly popular on the lake. As many as fifteen tip-ups are permitted, and live bait—preferably smelt—worked a few feet under the ice can produce big scores on lakers and the occasional salmon.

It is important to remember that an invisible line bisecting Lake Champlain is the boundary between New York and Vermont. Once you pass that boundary, you will need a valid Vermont fishing license.

Boat access to the lake is excellent, with the DEC operating seven hard-surface ramps in Washington, Essex, and Clinton Counties, and the State Parks Department operating another three in Clinton County. The DEC ramps are located in South Bay, Ticonderoga, Crown Point Reservation, Port Henry, Westport and Willsboro Bay, and Peru Dock. The Parks Department operates two ramps at Port Au Roche and a third at the Great Chazy River.

11 Lake Ontario, Western Basin

Lake Ontario is a great defining body of water for the state of New York. It serves essentially as the "roof" of the state, describing the northern boundaries of no fewer than seven counties (west to east, they are Niagara, Orleans, Monroe, Wayne, Cayuga, Oswego, and Jefferson). It also describes hundreds of miles of boundary between the United States and Canada. Most important to this work, however, Lake Ontario defines American cold-water sport fishing at its absolute best.

The lake generally is thought of as two basins: the Eastern Basin, which proceeds westward from the merge point with the Saint Lawrence River and including such fabled fishing venues as Sackets Harbor, Mexico Bay, and Sodus Bay; and the Western Basin, proceeding eastward from the Canadian border, the mouth of the Niagara River, Four Mile Creek, Wilson, and Olcott.

Both basins are massive in every regard, such that any attempt at brief description will fail. However, in highlighting a few features of these anglers' paradises, we can at least suggest the resources of great Lake Ontario.

Any consideration of fishing Lake Ontario almost has to begin with the remarkable size of the fish regularly taken here. As noted in *New York Sportsman* magazine, "Lake Ontario holds the state records for every

Figure 2 ■ Jamie Wu of Taiwan shows off his fine 32-pound king salmon, caught off Olcott in Lake Ontario in July 1998. *(Photo: Courtesy Niagara County Tourism.)*

trout and salmon species with the exception of two (brook trout and pink salmon)." State-record brown trout (33 pounds 2 ounces) and rainbow trout (26 pounds 15 ounces) were among the records set in the Western Basin.

The season for king (chinook) and coho (silver) salmon in the Western Basin is late June through September. The kings average 18 to 20 pounds with some running into the mid-40s. The cohos average 8 to 10 pounds

Lake Ontario's salmon and trout runs begin each year in the spring off the Niagara River in the Western Basin. Massive numbers of brown and rainbow trout and coho salmon run into the western tributaries, notably the lower Niagara River, Oak Orchard Creek, Johnson Creek, Sandy Creek, the Genesee River, Maxwell Creek, and the Irondequoit Bay. As the season progresses, salmon disperse eastward, and from mid-July on, the fishing can be good all along the middle and eastern reaches of the lake.

Superb shore fishing is available at hundreds of points in the Western Basin, especially including Four Mile Creek State Park, Twelve Mile

Creek, Tuscarora Bay, the piers at the foot of Route 425 in the village of Wilson, the Krull Park piers, Keg Creek and Golden Hill Creek State Parks, the pier at Oak Orchard Creek, the stone jetties at Hamlin Beach State Park, the west jetty at Sandy Creek, the stone jetties at Bradock Bay Park, the Genesee River piers, and the Irondequoit Bay outlet.

The big fish move off shore to cooler water in early summer as inshore waters exceed their comfort zone of 50 degrees. Then they can be taken by deep trolling and a new exciting season of fishing begins. Salmon in the 20- to 40-pound range and massive steelhead are regularly caught in late spring and summer by downrigger-equipped boats.

The DEC's program for stocking Lake Ontario and key tributaries is one of the most scientifically managed in the country, and one of the most dramatic. In 1998 3.5 million trout and salmon, including 1.6 million chinook salmon, 245,000 coho salmon, 357,900 brown trout, and 417,000 lake trout, were stocked directly in the lake or into tributary streams. Embayments were stocked with 104,480 walleyes.

Tip: Top baits include fresh clumps of egg skein, egg sacks, egg imitations, minnows, or Kwikfish lures. You want a vertical presentation to minimize snags and maintain soft contact with the bottom. Using a rig involving a three-way swivel, a 1- to 2-ounce pencil lead and 6- to 8-foot leads off the trailing eye is the most popular way for taking (steelhead) trout in the river.

From early spring through December, the Western Basin offers a spectacular diversity of sport and recreational fishing. The mouths of feeder "streams" in Niagara County, including Twelve Mile Creek, Eighteen Mile Creek, and the Lower Niagara River are alive with smallmouth bass and walleyes in spring and early summer. Huge lake trout, steelhead and northern pike also ply these fish-rich waters, as do muskellunge, pickerel, and big crappie.

The size of individual fish is enough to lure the most jaded angler to Lake Ontario's Western Basin. However, what is truly astounding is the opportunity to catch not one huge fish but one huge fish after another on any given day.

12 Lake Ontario, Eastern Basin

Directions: Route 104 parallels the shoreline of Lake Ontario its entire length in New York, across all seven counties from Niagara to Jefferson.

The Eastern Basin of Lake Ontario is considered by many anglers to be among the finest freshwater fisheries in the world.

Salmon, trout, steelhead, and walleye thrive and grow to mind-boggling sizes here. Consider some of the records. In July 1998, the standing state records set in the Eastern Basin—in the lake or the mouths of tributaries to the lake like the Salmon River—included chinook salmon (47 pounds 13 ounces), coho salmon (33 pounds 4 ounces), and Atlantic salmon (24 pounds 15 ounces). These records continue to be broken—only to be replaced by even larger Lake Ontario fish. For example, in January 1999, a 33-pound 7-ounce coho was caught by angler Stephen M. Sheets, Jr. fishing out of Oswego Marina. The fish, currently in the confirmation process with the International Game Fish Association, is thought to be a new world record. And in April of 1999, a new state-record 25-pound 15-ounce Atlantic salmon was caught offshore near B. Forman Park in Wayne County by a twelve-year-old angler named Mike Dandino. In addition, experts predict that the next world record for walleye will be taken where Eastern Lake Ontario flows into the St. Lawrence River. In fact, the waters of Eastern Lake Ontario were chosen by *The Fisherman* magazine as the number-one location in North America to catch trophy walleye tipping the scales at 8 to 12 pounds each.

April is the traditional beginning of the season as salmon move eastward out of the Western Basin. In addition to salmon, this is the time many oversized browns and lake trout are caught. May and early June are prime times for walleyes and steelheads as they make their inland runs. Enormous walleyes are caught in the Black River Bay and giant steelheads are taken below the Black River Dam, in Henderson Harbor, and in the mouths of North and South Sandy creeks.

Henderson Harbor is an especially popular location in the Eastern Basin. It is sheltered water, so anglers driven off the lake by bad weather have the pleasant option of catching big brown trout and smallmouth bass here. For boats fishing out of this harbor, walleyes, huge lake trout, and big brown and rainbow trout are definite possibilities. But most are after big king salmon. Most big fish are taken by trolling, either plugs or spoons. Manny has a soft spot for Henderson Harbor ever since the day he steamed out on the *Jessie May* and hooked the largest fish he ever caught while trolling anywhere. It was a 29-pound king salmon and it tore off line like a fast freight train—an unforgettable experience.

Sodus Bay is another great spot, both for fishing and for the fleet it harbors. Brown trout, smallmouth bass, northern pike, and big crappie (locally called "strawberry bass") and all three varieties of salmon are taken in Sodus Bay, and boats out of Sodus regularly return with trophy fish. Salmon runs in the mouths of the major rivers in this region, especially the Salmon, the Black, and the Oswego, can be extraordinary. (*Sports Afield* magazine described the area within 300 yards of the Salmon River breakwater as being one of the two best places to fish in all North America.) Significantly, steelhead follow the spawning salmon, feasting on eggs. All of these major rivers have steelhead populations much of the year, but they are found in greatest numbers in fall and winter.

During the summer, Eastern Lake Ontario offers some of the best deep-water fishing for salmon and trout in North America. In June and July you can enjoy great small and largemouth black bass action, too.

Tip: Captain Ernie Lantiegne, guide and fishery biologist with DEC, notes that versatility catches fish. Learn to use all types of fishing gear, he advises, and you'll be more successful on Lake Ontario. Don't just follow the crowd. Wire line, led-core line, or drop-sinker rigs are extremely effective, as are diving planers, fished from the boat or from downriggers. Also, don't get hung up on a standard trolling setup. If a system is working, don't change it. If things are slow, try running longer or shorter lengths of line to your lures. In crystal clear water, if things aren't happening, try reducing the amount of gear, especially downriggers in the water.

Running from the end of the summer (around Labor Day) through late November, the tributaries up and down the lakeshore are filled with trophy salmon—king, coho, and Atlantic. In Jefferson County, for example, at the lake's eastern edge, North and South Sandy Creeks receive incredible runs of chinook salmon. Black River, the primary Lake Ontario tributary stocked with Atlantic salmon, gets a phenomenal Atlantic salmon run.

With the onset of colder weather, fishing for steelhead remains very good close to shore. Throughout the winter you'll find unbeatable ice fishing for northern pike and yellow perch in many bays and marshes, especially in Sodus Bay.

It doesn't get any better than this.

13 Lake Erie, Eastern Basin

Directions: N.Y. Route 5 parallels most of Lake Erie's New York shoreline, from Buffalo to the Pennsylvania border. U.S. Route 190 traces a portion of the lake north of Buffalo.

At 241 miles long and 57 miles at its widest point, Lake Erie is the eleventh largest lake in the world and fourth largest of the Great Lakes. To understand Lake Erie is to see it as three distinct "basins." The Western Basin, which is the shallowest, extends from the western end of the lake to about Cedar Point, Ohio. The Central Basin extends to the edge of the trenches in Erie, Pennsylvania. The Eastern Basin extends from Erie to the eastern edge of the lake in Buffalo, New York. All of New York's Lake Erie waters are located within this Eastern Basin, which is the deepest of the three. Because of its depth, its great feeder streams, its forage fish resources, and its widely varying habitats, the Eastern Basin is an exceptionally diverse and productive fishery.

Perhaps the most abundant species in all three basins is the walleye, which originate from a number of wild populations that spawn in rivers or on lake shoals featuring gravel- or rubble-strewn bottoms. A small fraction of Western Basin walleye actually migrate 200 miles or more from the Western to the Eastern Basin to spawn. In summer, owing in small part to this migration, thousands of large female walleye of 5 pounds or more are taken in New York waters, with "eyes" in the 7-pound class not at all unusual—especially at night.

Tip: One of the truly exciting experiences when night fishing for walleye is the frantic surface splashing of terrified forage fish herded together by packs of hungry walleye. When you hear this splashing, quickly hook on a fresh (flat-lined) bait fish or tie on crankbaits that closely imitate shad, alewife herring, or small yellow perch, cast into the disturbance and fasten your figurative seatbelt!

The second most widely sought species in Lake Erie is the smallmouth bass, and with good reason. Large populations of smallmouths ply hundreds of rocky points, rock faces, drop-offs, and creek mouths in the Eastern Basin, and bass anglers enjoy high fishing success rates, with a surprising number of bronzebacks taken in the 18- to 20-inch class. The state-record smallmouth, a whopping 8-pound 4-ounce battler, was taken on a jig and grub tail in Lake Erie by Andrew C. Kartesz in June of 1995. (Of interest: longstanding state-record smallmouths in Pennsylvania [7 pounds 10 ounces] and Ohio [9 pounds 8 ounces] were taken in Lake Erie.)

Smallmouth fishing is especially productive in early spring near the mouths of such key feeder streams as Eighteen Mile Creek, Cattaraugus Creek, and Chautauqua Creek. Most smallmouths tend to suspend in slightly deeper water (15 to 30 feet) in later spring. Great locations for smallmouths include the lake's many harbors, including Buffalo Boat Harbor, Dunkirk Harbor, and Barcelona Harbor.

The lake's cold-water species, although less publicized, also provide sensational fishing opportunities, largely courtesy of the DEC. For example, lake trout stocks had declined because of a burgeoning population of sea lampreys. However, the DEC's sea-lamprey control program was so successful that plenty of lakers are being caught again, with fish in the 20-pound class being taken. Moreover, the DEC's continuing stocking program—aimed at developing naturally reproducing populations—promises even better and larger sustaining stocks.

Coho salmon, chinook salmon, rainbow/steelhead trout, and brown trout also have been liberally stocked in the lake and its tributaries over the past two decades. Of these nonnative populations, the rainbow/steelheads are the most successful, providing truly great fishing in the lake and its

Figure 3 ■ Eleven-year-old Julie Muldoon caught this beautiful 5-pound 7-ounce smallmouth drifting worms with her granddad in Lake Erie.
(Photo: Julie Muldoon.)

feeder streams. However, the lake also provides high-quality salmon angling opportunities that are not to be missed. The longstanding New York record pink salmon (4 pounds 15 ounces) was taken in Lake Erie in 1985 by Randy Nyberg, and the state-record yellow perch, a 3-pound 8-ounce buster, was taken in the lake three years earlier by George Boice.

Boat access is excellent, with launch sites scattered from the Pennsylvania to the Canadian borders. Three relatively new and exceptionally convenient public launch facilities built by the DEC and local authorities are located at Sturgeon Point Marina, Cattaraugus Creek Harbor, and Barcelona Harbor. An especially interesting shore-fishing structure, featuring year-round fishing, is to be found at the Niagara Mohawk warm-water discharge point at Dunkirk Harbor.

The Finger Lakes

The Finger Lakes Region of New York is a paradise of outdoor beauty and fishing opportunity. All of its eleven lakes are glacial. The largest are Canandaigua, Keuka, Seneca, Cayuga, Owasco, and Skaneateles; the smaller are Conesus, Hemlock, Canadice, Honeoye, and Otisco. Together they offer more than 650 miles of shoreline and often spectacular opportunities for walleye, pike, bass, tiger muskies, lake trout, rainbow and brown trout.

14 Conesus and Honeoye Lakes

Directions: Take Route 20A in Livingston County to the juncture with East Lake Road at the northern tip of Conesus Lake; follow East Lake the length of the lake down to the Conesus Inlet Wildlife Management Area at the southern tip. For Honeoye, the directions are the same. Take 20A in nearby Ontario County to East Lake Road; then follow East Lake along the lake.

Conesus and Honeoye are two of the smaller Finger Lakes that are so similar in fishing conditions that they are conveniently discussed together. Both are warm-water fisheries, with abundant populations of largemouth and smallmouth bass, walleye, and pike. In fact, Honeoye is the site of annual bass fishing tournaments, where largemouths of 7 pounds or more have been taken.

While anglers on the other Finger Lakes tend to be attracted by cold-water species such as trout or salmon, those knowledgeable in the ways of bass fishing detour around to Conesus or Honeoye. A particularly devoted partisan named Pat Smith once wrote a paragraph that helps explain the devotion of bass enthusiasts. "But if the salmon and trout must be classified as elite in this mythical social structure, then let the black bass be given permanent status as the working class of American game fish. He's tough and he knows it. He's a factory worker, truck driver, wild catter, lumberjack, barroom bounder, dock walloper, migrant farmhand and bear wrassler. And if it's a fight you're looking for, he'll oblige you anytime, anywhere. Whether it's a backwater at noon, a swamp at midnight or dockside at dawn, he'll be there waiting."

Plant life in both lakes includes eelgrass, pondweed, water stargrass, and Eurasian milfoil, from the weedy edges and bays out to depths of 15 feet or more. Consequently the habitat for cover, feeding, and breeding is excellent for largemouth bass and pike. In addition, both lakes feature large expanses of rocky bottom and gravel where smallmouths congregate. Rocky shoals, drop-offs, and points yield fine catches of bronzebacks, especially to anglers skilled in the presentation of natural baits such as crayfish and minnows.

Both lakes are superb walleye fisheries, primarily because of scientific stocking programs. Walleye are the only fish stocked in Honeoye. In a program calculated to keep a high density of predators to control the plentiful panfish populations, DEC stocks about 8.7 million walleye fry annually in this lake. Larger fingerling walleye (65,000 in 1998) are stocked every year in Conesus. One favorite method of catching walleye is trolling crankbaits, especially during low light conditions (early morning, evening, and night.)

Tip: A key to trolling success for walleye is working the right depth. Walleye trollers should have a selection of shallow, medium, and deep diving lures at their disposal. When the graph marks fish at a certain level—at 15 feet, for example, or at 40 feet—the angler can select lures that will put the bait right where the fish are holding. Remember, line diameter and length of line behind the boat affect how deep a crankbait will dive.

Figure 4 ■ Ken Kierst smashed the New York State black crappie record with this 3-pound 12-ounce bruiser.
(Photo: Pete Kierst.)

The DEC also stocks tiger muskellunge in Conesus, and the occasional monster is taken on large, noisy lures and large live bait. There are also excellent populations of yellow perch, bluegills, rockbass, and crappie in both lakes. At one and the same time, these provide good forage for the game fish and super action for anglers out to fill a stringer with panfish.

Boat access to both lakes is excellent, including an off-public-road, hard-surface ramp 4 miles south of the hamlet of Honeoye off E. Lake Road, and a hand-launch site for car-tops off Sandy Bottom Road. Four launch sites on Conesus Lake include a hard-surface ramp off E. Lake Road and three hand-launch sites at the south end off Route 256, at Sand Point off Route 20A and Pebble Beach Road, .25 miles south of 20A.

15 Keuka Lake

Directions: Take U.S. 390 to the Route 54 turnoff in Yates County. Follow Route 54 along the entire eastern shoreline of the lake and Route 54 along the western shore.

Keuka Lake was a fishing paradise to the native Americans long before the first settlers reached America; a beautiful and bountiful lake to be honored and cherished.

Today Keuka Lake still holds a place of honor with anglers who are wise to its opportunities. The 11,730-acre lake has over 60 miles of shoreline and a maximum depth of 186 feet. It was carved by glaciers into the shape of a "Y," and each of its three arms is just large enough to be explored and fished in a day by an enterprising angler. Significantly, each arm is relatively slender, with good protection from sheltering shorelines. Consequently, wind is seldom a serious problem to offshore anglers, many of whom elect to drift or troll at a controlled pace.

Most anglers come to Keuka with trout or bass in mind. The lake is an excellent brown trout fishery; indeed, a 23-pound 12-ounce brownie taken in Keuka some years ago was at the time a state record. However, the lake also ranks among the best of the Finger Lakes for lake trout, smallmouth bass, pickerel, and yellow perch.

Perhaps most important, Keuka is one of only four Finger Lakes (the others are Hemlock Lake, Seneca Lake, and Cayuga Lake) to boast a good land-locked salmon fishery; a fishery that New York State works hard to preserve. In a recent stocking season, the DEC stocked Keuka Lake with more than 22,000 land-locked salmon, in addition to 110,000 brown trout.

Ice out is the traditional time to surface troll for land-locks, usually at slower speeds until the water temperatures hit 45 degrees. (Even in warmer waters, however, trollers often hit pay dirt at or near the surface on days when the water is rough and the skies are overcast.) In warmer weather, the preferred method for taking salmon is fast trolling at various depths, depending upon temperature and conditions.

Deepwater trolling is less affected by weather conditions but speed is still critical. The key to deep trolling success is more often speed than a particular lure, although familiar silver-finished flutter spoons seem to produce the best results. (Keuka Lake salmon feed on sawbellies as well as the more traditional diet of smelt; consequently lures do not necessarily have to mimic smelt in shape and flash.) In the fall, gaudier lures appear to be particularly appealing to salmon, especially when trolled tight to windward shores near the mouths of salmon-spawning streams.

Tip: Knowing what a fish prefers to eat at all given times of the year and presenting artificial lures and live bait in the most natural fashion are vitally important to catching fish consistently. If you haven't fished this lake or other waters before, do your homework. Call and visit a few bait-shop owners to learn everything you can about the lake or river. Study maps. Consider hiring a guide for the first day out on the water. Get ready to succeed.

Keuka's big lake trout are generally found in deep water and are most efficiently caught on live sawbellies fished at or near bottom. For anglers who may look down on this type of fishing, we commend the statement by fishing writer Louis D. Rubin, Jr., who once said, "To a far greater degree than other kinds of fishing for pleasure, the art of bottom fishing involves the actual catching of fish!"

Local outdoor columnist John Collum notes that "Keuka is one of the most fickle of all the Finger Lakes. One day it will offer the fastest fishing found anywhere and the next day yield nothing." This interesting challenge makes the lake all the more attractive to us.

Keuka is one of the most accessible of the Finger Lakes, with developed highways paralleling virtually every foot of both shorelines. Hard-surface boat ramps are maintained at the lake's northern tip in the Village of Penn Yan and in Keuka Lake State Park off Route 54A on West Bluff Road.

16 Seneca Lake

Directions: Take Interstate 90 to exit 42, halfway between Syracuse and Rochester. Follow signs to Route 14, which traces the entire western shore of Seneca Lake.

The deepest and second longest of the Finger Lakes is Seneca Lake, which comprises some 43,800 acres. Depths exceeding 600 feet provide wonderful habitat for trout, especially lake trout, which account for more angling activity on this gorgeous body of water than any other fish. Lake trout prefer water from 48 to 52 degrees, colder than other game fish prefer. During summers, lakers characteristically descend to 200 feet or more in search of cold water. Clearly they can find any depth that is comfortable to them, regardless of season or weather, in this exceptional lake.

Lake trout spawn in Seneca Lake in the fall; although the DEC regularly stocks lakers (100,000 fish in one recent stocking season, ranging from 6 to 8.5 inches in length) in the lake, most of the fish taken are wild trout.

It is thought that Seneca Lake might very well produce the state record laker one of these days, replacing the current 39-pound 8-ounce monster taken in Lake Ontario in 1994. Fred J. Kane, an authority on the lake, notes that the east side of the lake is most productive. Trolling downriggers is an effective way of catching lakers deep, with the Seth Green trolling rig and a wobbling spoon or plug tied to each leader a local favorite. Another popular method is fishing beneath lanterns and an anchored boat in

Figure 5 ■ Daniel Miller caught this fine yellow perch in Seneca Lake casting a brown bucktail jig he made himself.
(Photo: New York State DEC.)

deep water with live sawbellies suspended at mid-depth. The light attracts baitfish and the baitfish attract lakers.

> *Tip:* *Our own preferred method for catching lakers is fishing from a double-anchored boat on a good-looking bottom slope in 80 to 120 feet of water, using alewife herring either suspended, just off bottom or "Uncle Nick" style, in which the sliding sinker rests on bottom, allowing a 4-inch shiner to work upwards to the length of the leader. Our preferred rig is a 1-ounce sinker above a black swivel, with about 3 feet of leader and a #6 hook.*

Brown trout also are stocked in the lake each year at Lodi, usually fish ranging upwards to 8.5 inches in length.

However, rainbow trout provide much of the allure of this great fishery, not least because of the record-threatening fish taken each spring from Catherine's Creek, which feeds Seneca Lake. Rainbows of 20 pounds or more are taken in this area each year. The rainbows are a mid-water species, often found offshore feeding at the top of the thermocline. Many "bows" are taken by trolling Seneca Lake using spoons, plugs, or spinners with downriggers, diving planes, or wire line. However, natural baits, including herring, minnows, and worms, also take their share of rainbows.

Smallmouth and largemouth bass also teem in the lake, the former along the lake's plentiful rocky points and outcroppings; the latter in the weedier, shallower waters. Bass weighing upwards of 6 pounds are taken by trolling, drifting, or casting. Northern pike and yellow perch are also found in shallower waters in the lake in very considerable numbers.

Regarding access, landowner permission must be requested in certain portions of the lake. However, excellent access is available along long stretches and in the state parks situated on the east side of the lake and in the Willard Wildlife Management Area on the opposite shore. In addition, boat launches with hard-surface ramps are located just east of Geneva along Routes 5 and 20, in Seneca Lake State Park, and off Route 96A north of Willard in Sampson State Park.

17 Cayuga Lake

Directions: The lake lies roughly equidistant between Rochester, Syracuse, and Binghamton, and touches the city of Ithaca on its southern tip. Take Route 13 to the Cayuga Inlet at Ithaca; follow Route 89 along the entire western bank of the lake to its Seneca River outlet.

When ESPN considers sites to film fishing shows, its producers often think of Cayuga Lake. In 1998 alone, this premier sports channel filmed several bass fishing episodes on the lake, and leading television fishing personalities Lonnie Stanley and Jimmy Houston filmed shows of their own. *Outdoor Life* magazine named Cayuga Lake one of the top ten bass lakes in the United States in 1998. During the season, there is a serious bass tournament on the lake every weekend. And this only begins to describe the fishing here.

Cayuga Lake is a long, slender glacial lake that extends for some 40 miles, with about 100 miles of shoreline lying in Seneca, Cayuga, and Tompkins Counties. Fed by four feeder streams near Ithaca, the lake comprises nearly 43,000 acres. Perhaps its most prominent feature is its depth, running down to 435 feet at its deepest point at the mouth of Shell-drake Creek.

Cayuga is a great bass lake. Teacher and Cayuga Lake guide Paul Tatar has taken smallmouths up to 4.5 pounds, and some largemouths are taken in sizes generally associated with South Georgia and North Florida.

Tip: As Paul Tatar notes, in Cayuga you don't have a great deal of the traditional kinds of structure where largemouths and smallmouths lie. In spring, there are a few points and docks to work, but once warmer weather sets in, you'd better learn to jig in deeper water. The ideal way to catch summertime smallmouths is to locate a hole off of a large point in deep water and jig spoons. The lake is full of bass, but you have to be creative and thoughtful to find and catch them.

As its depth suggests, Cayuga also provides fine habitat for cold-water salmonids, especially trout and even more especially lake trout. In addition, the lake is a fine fishery for land-locked salmon, thanks in no small part to stocking efforts by the DEC. In a recent stocking season, some 185,000 land-locked salmon and brown, rainbow, and lake trout were stocked in the lake at Ithaca. (Significantly, the salmon now being stocked are as large as 12 inches to discourage the numerous lake trout from gobbling them up.)

In the fall, salmon and brown trout run in from deep drop-offs to shallow water and up the streams to spawn. (Local anglers say the best day of the year for salmon is Election Day.) When temperatures run from cool to cold, anglers catch fine salmon and trout trolling such stickbaits as Rapalas and Rebels on or near the surface. In warmer weather, downriggers are needed to get down to the thermocline, which can vary from 35 to 120 feet.

Depths drop off from 20 feet to 200 feet in less than three-quarters of a mile at the southern tip of the lake. This creates the kind of slopes and other irregular bottom features that are ideal for locating lakers. And in Cauyga Lake, the lake trout fishing can be extraordinarily rewarding, with fish in the 12- to 15-pound range caught each year.

Access to the lake is excellent, with roads closely paralleling both shorelines. Boats and motors of any size are permitted, and four state parks with ramps make for easy launching. Developed ramps for larger boats are located at the north end at Long Point State Park and Cayuga Lake State Park and at the south end in Allan Treman Marina. A launch for small boats is located 10 miles up the west side of the lake in Taughannock State Park.

18 Owasco Lake

Directions: Follow Route 38 south of Auburn in Cayuga County along the entire 11-mile length of the lake's shoreline.

Owasco is one of the smaller Eastern Finger Lakes at some 6,600 surface acres. Long and deep, the lake is beautiful in all of its aspects, and most especially beautiful to anglers intent on a great day of northern pike, bass, lake trout, or brown trout action. The water in the lake is remarkably clear and runs to depths of 170 feet in a large area almost exactly at the lake's center.

Owasco is noted for its trout fishery; indeed it is commonly thought to have the best trout fishing of the small Finger Lakes. Each year, a number of trophy browns weighing in double digits are taken, and rainbows, although smaller, are caught in significant numbers as well. As the deep water might suggest, Owasco also is a superb lake trout fishery year-round, with particularly hot action in May and June. Lakers prefer water from 48 to 52 degrees, colder than any other game fish prefer, and during warmer weather, they will descend as deep as necessary to find cold water. Unlike other trout species, lakers spawn in lakes rather than moving water. Spawning occurs in fall over rocky bottoms in depths that can vary from 5 to 20 feet or more.

In Owasco, lake trout move into shallow water in the spring and again just before the fall spawning period. They feed almost exclusively during the day, being sight feeders. The two favorite methods of catching spring fish are casting flashy spoons or still fishing with live bait. In warmer weather, most anglers favor downrigger trolling to get the bait deep and keep it there. This is especially popular in Owasco, where boats and motors of any size are permitted. Certainly skilled trollers can score well. However, we are not persuaded that a carefully double-anchored boat over just the right drop-off and live bait fished deep will not put fish into the net as productively if not more productively than trolling.

Tip: Many lakers are killed unnecessarily when excited anglers reel them up too quickly from deep to shallow water. The fish are affected with a malady very much like "the bends" and don't survive, even if released carefully. Remember when fishing in deep water to work your fish up at a pace that will give it every chance to live to fight another day if you want to release it.

The lake and its tributaries, prominently Owasco Inlet, are heavily stocked with trout. Last year, DEC placed 14,450 browns, 25,000 rainbows, and 16,000 lakers in the two bodies of the lake and its principal feeder stream. At the same time, the Owasco Lake Anglers Association stocked 124,000 walleye fingerlings, with the clear intention of sparking a meaningful walleye fishery in the lake. Challenges in stocking smaller fish, however, are the already well-established northern pike and smallmouth bass populations, which feed voraciously on fingerlings and fry.

Not surprisingly, the combination of fish species and latitude of Owasco makes for fine ice fishing, especially for trout, bass, pike, and panfish. We urge caution to those hardy enough to venture out on the ice, those whom Arthur R. MacDougall, Jr. refers to as "having gotten fed up with being comfortable and sane." Thick ice, layered clothing, good gloves, and hand warmers are minimum requisites, along with the various and sundry items of equipment (ice auger, tip-ups, etc.) that accouter this special brand of sport.

Boat access is available at both ends of the lake, with paved public launch facilities maintained at Emerson Park at the north end of the lake. Shore access also is available at Emerson Park. When planning a trip to the lake, the Owasco Inlet should not be overlooked, especially upstream of Moravia, where many deep pools provide truly fine fishing for rainbow and brown trout and the state provides plenty of public fishing rights.

19 Skaneateles Lake

Directions: Take Route 20 in Onandaga County to the village of Skaneateles; turn on Route 41, which follows the entire eastern side of the lake to its tip in Cortland County. Route 41A traces the top half of the western shoreline.

Skaneateles has been called the "real sleeper" among the Finger Lakes for outstanding fishing quality. Often overshadowed by the reputations of nearby Seneca and Cayuga Lakes—both significantly larger and better known—Skaneateles is itself a superb fishery, especially for trout and land-locked salmon.

Skaneateles is the highest of the major Finger Lakes and the clear-est, primarily because it is spring-fed via six inlets (Grout Brook, Bear Swamp Creek, Harold Brook, Shotwell Brook, Five Mile Brook, and Ten Mile Brook) along both sides and at the southern end of the lake. The cool, clear water from these inlets keeps the water purity excellent and the levels of oxygen good for fish habitat year-round.

The lake's name comes from the Indian word Skan-e-a-dice, meaning "long lake," and this certainly is descriptive. From Glen Haven Inlet at the southern end to the Skaneateles Creek outlet at the northern end, the lake is 15 miles long, with an average width of .9 mile.

Skaneateles provides fishing to satisfy almost any interest and incli-nation. Anglers find fabulous fishing for brown and rainbow trout, espe-cially near shore in early spring and late fall. Trolling with small spoons and streamers accounts for many of the trout taken in the lake in spring, although casting from shore with bait (minnows or "popped-up" combi-nations of marshmallows and worms) is also very effective. Grout Brook, one of the lake's key feeder streams, also merits special attention in early spring, when many large spring-run rainbows are taken.

Tip: Troll silver-plated lures along the drop-offs from Pine Grove to the New York State launch area for big browns, "bows," and wild-strain lake trout when the weather and water temperature are a bit warmer.

The lake is deceptively deep; in fact, with depths running to 350 feet, it is the third deepest of the Finger Lakes, with only Seneca and Cayuga

Lakes having deeper points. This depth is significant for all of the cold-water species and most especially for lake trout, which reproduce in the lake. On average, the lakers tend to run a bit smaller (15 to 20 inches) here, but the catch rate in most seasons is better in Skaneateles than in any other Finger Lake.

A thriving land-locked salmon fishery also is developing in this clear, beautiful lake, courtesy of the DEC. Scientific stocking programs have added this exciting species to the already fine cold-water populations. In one recent stocking season, for example, 5,000 land-locked salmon 6.5 inches long were stocked directly into the lake, along with more than 27,000 rainbow trout.

Skaneateles also is excellent for smallmouth bass, largemouth bass, pickerel, and panfish, especially in the shallower water at its northern and southern ends. The whole spectrum of stickbaits, spinnerbaits, crankbaits, and live baits put bass and panfish on stringers. Minnows are especially effective for pickerel; crayfish are deadliest for smallmouths.

Boat access to the lake is provided via a concrete ramp off Route 41A, 2.5 miles south of the village of Skaneateles.

Skaneateles is its own kind of perfection: a beautiful, clean, uncrowded lake, jammed with fish of all descriptions, awaiting fly or spinning lure or plug or bait, fished from boat or shore. It calls to mind the lines of author Russell Chatham, who wrote: "Fishing with a fly, bait, a handline; I don't much care. Fishing, in my estimation, is not a hobby, a diversion, a pastime, a sport, an interest, a challenge, or an escape. It is a necessary passion."

Major Rivers

Like much that is great in New York State, the Great Rivers—the Hudson, the Delaware, the Niagara, the Salmon, the Mohawk, and the Susquehanna—are impressive by their size, their richness, and their power. The lower Hudson is said to be the greatest natural wildlife resource in the state. The massive St. Lawrence flows 800 miles, from the Canadian border to the Atlantic. The Niagara plunges 185 feet in one of the most spectacular natural phenomena in the nation. And for all this muscle, the great wild rivers of New York offer endless pleasure and contentment for anglers who know how to fish them.

20 St. Lawrence River/Thousand Islands Region

Directions: Take I-81 North to the Thousand Islands bridge; turn left at exit 50 onto Rt. 12, which parallels the river in the Thousand Islands region from Fishers Landing to Cape Vincent.

The St. Lawrence River runs 800 miles from Lake Ontario to the Atlantic Ocean and serves as a border between the U.S. and the province of Ontario for 114 miles before flowing entirely into Canada. As its geography would suggest, this great river is a spectacular fishery along its entire length. However, no section of the river is more famous or more productive than the

40 miles of headwaters region between Cape Vincent (at the mouth of Lake Ontario) and Alexandria Bay. This is the famed Thousand Islands region, where the world record muskellunge—a 69-pound 15-ounce monster—was boated by an angler named Arthur Lawton more than three decades ago.

This region still boasts one of the two greatest populations of natural (unstocked) muskies to swim anywhere in Americas. The fish average 12 to 30 pounds and an occasional fish in the 40- to 50-pound class is taken. The best muskellunge fishing is in October, when these brutes are feeding heavily in anticipation of winter. Captain Myrle Bauer, an expert on the Thousand Islands fishery, has great success trolling for big muskies. One of the best fish taken by his boat was a 50-pound beast caught and released by his wife Beth.

What lures most anglers to these waters is the incredible richness and variety of the fishery, including muskies, certainly, but also fabulous bass, pike, and walleye. *Bassmaster* selected the Thousand Islands region as one of the ten top destinations in the nation, noting that "few places in the world offer the variety of abundant game fish as does this part of the St. Lawrence." The authoritative bass magazine added, "On a summer day, it's not unusual to catch 20 or 30 smallmouths, 15 or so largemouths, two dozen big (5 to 8 pounds) hard-charging pike, and a boatload of tasty walleye. But for the serious bass fisherman, the area's mixed bag of plentiful largemouths and smallmouths is the main attraction."

The clear, cool water rushing through rocky channels, over drop-offs and shoals, through weedbeds and quiet bays provides such ideal bass habitat that the Bass Masters Classic tournament has come to the region twelve times in the past two decades. Anglers working these waters are confronted with a choice: whether to work the piles of rocky rubble, drop-offs, points, and deeper holes for bronzebacks, or to ply the abundant weedbeds and cattail-filled shallows in search of largemouths. Happily, it is almost impossible to make a bad choice. And for variety, Lake Ontario and its bass-rich Chaumont and Sodus Bays are but a bass boat ride away.

Smallmouth average 10 to 15 inches in length, with fish up to 3 pounds commonplace and monsters over 4 pounds not unheard of. The largemouths are larger if a bit less plentiful. Drifting live bait is perhaps the most popular method of taking fish, followed closely by spinner baits enhanced with night crawlers or with white, yellow, black, or chartreuse skirts, and plugs that imitate shiners. Jigs with plastic tails work well when fishing deeper (the river runs to almost 300 feet in spots) for smallmouths.

Big northern pike also make the Thousand Islands area a first-rate fishing venue. In fact, the northerns are so plentiful that they sometimes complicate things for touring bass pros. When water temperatures are still in the 50s, pike are found in the shallower reaches, feeding aggressively on the schools of alewife herring that abound at this time of the season. Drifting shiners through the weedbeds is the most effective method of taking big northerns. This is a technique that Ron and good buddy Bob Findley have worked to perfect when pike fishing each year just north of the Thousand Islands region. Whereas some local guides prefer double-hooking a shiner—that is, running the hook and line through the shiner's mouth, then placing the hook just under the skin at the back of the dorsal fin—we prefer a more lifelike presentation, using a weedless #6 hook worked up through the shiner's upper lip and out one nostril. Pike in the Thousand Islands region grow large and mean. A number of lunkers from 15 to 20 pounds are caught here each year.

Tip: A straight monofilament (8- to 10-pound test) leader will produce many more strikes than a wire leader in the ultraclear Thousand Islands waters, whether using live bait or artificials. While a pike's teeth are sharp, they are more like needles than razors; thus abrasion is more likely to sever your leader than the fish biting through it. Check your leader and be prepared to change it after each pike strike.

The walleye population has grown in the last few years as a result of reduced alewife populations (because alewives eat walleye fry.) Some of these fish have grown to record-menacing size. Seven- to 8-pounders are commonplace, and fish in double numbers are taken each year. In the spring, walleyes are found in the same haunts (and taken with some of the same lures/baits) as largemouth bass. Rapalas and small crankbaits tossed to the edges of weedlines in shallow water—especially in overcast conditions—produce walleyes. As the water warms, walleyes move out into deeper water and are caught along the ledges and drop-offs on jigs and nightcrawlers. The mouth of the Oswegatchie River is one of dozens of prime walleye spots.

Of all destinations along the Thousand Islands area, Cape Vincent is the largest and best-disposed to angling. Located where the St. Lawrence begins and the Eastern Basin of Lake Ontario ends, Cape Vincent provides a handful of marinas and other boat-launching facilities, camping facilities, and plenty of fishing guides who are expert in finding big game fish in these waters.

We often comment on the natural beauty of this lake or that stream, and always with reason. However, it is likely that no view in New York is more spectacular nor more panoramic than the view from the Thousand Islands Bridge; a view that tugs equally at the heart and the resolve to head back to home and work after a St. Lawrence River fishing trip.

21 Niagara River, Upper

Directions: Follow the Robert Moses Parkway in Niagara County along most of the shoreline of the Upper Niagara.

The Niagara River is a 40-mile sportfish superhighway between two of America's greatest sports fisheries: Lake Erie and Lake Ontario. It is therefore not surprising that the river is itself great, in the fullest sense of the word.

The river is actually two distinct fisheries; one above Niagara Falls, one below. Above the falls, the river stretches from the city of Niagara Falls to the city of Buffalo, including two branches that curl around Grand Island. With an average depth of 25 feet, the upper river is essentially a warmwater fishery.

The upper river is one of the top producers of muskellunge in the world. These largest members of the pike family reproduce naturally in the Niagara and often tip the scales at the 40-pound mark. In the fall, three hot spots for muskies are the areas on either side of Strawberry Island,

Figure 6 ■ Terry Jones took this trophy 6-pound smallmouth bass at the mouth of the Niagara River on a Kalin Tube Jig. *(Photo: Niagara County Tourism.)*

the area just outside the Erie Basin Marine and the warm-water discharge of the Huntley Power station in Tonawanda.

Massive trout also are taken in the upper river: browns between 10 and 15 pounds and rainbows of 6 to 10 pounds are considered commonplace. The best timeframe for trout in the river is December through March. Fresh trout eggs are the favored natural bait; the Kwikfish lure is the local favorite artificial.

Tip: In winter, try for steelheads just above the upper rapids area along the Robert Moses Parkway and on the back side of Goat Island. Cast spoons, spinners, or egg sacks out into the current.

Largemouth and smallmouth bass, muskellunge, walleye, northern pike, perch, crappie, and a variety of other fish species are available throughout the year. Largemouths average between 2 and 3 pounds and trend up to 6 pounds. Smallmouths run a bit smaller but are so numerous that serious trout and salmon anglers sometimes consider them a nuisance. (In all but these extraordinary waters, a 2-pound smallmouth is highly prized!)

Yellow perch abound in the upper river, especially in the harbors and around piers. These panfish provide great sport, especially for young anglers more interested in action than records, and are great table fare. Given the healthy populations of yellow perch, it is not surprising that big walleye also are taken in the upper river, especially in the area around Riverside.

Big northern pike also cruise the weedbeds and shallows of the upper river, averaging more than 4 pounds but running as large as 20 pounds. These voracious fish readily take artificials that imitate forage fish, but prefer live bait—especially big minnows.

Drifting, trolling and still fishing are key approaches to taking trophy fish in the upper river. However, excellent quality fishing also is available from literally hundreds of points along the shoreline. Perhaps the most unusual of these is the covered fishing pavilion in Fisherman's Park in the City of North Tonawanda.

Boat access also is good in the upper river, with multiple launch ramps in Griffon Park and in Riverside Park in the city of North Tonawanda. In addition, the state maintains a concrete ramp launch facility at Big Six Mile Creek State Park Marina on the western shore of Grand Island.

22 Niagara River, Lower

Directions: Like the upper Niagara, the lower river is paralleled for its entire length by the Robert Moses Parkway.

The lower Niagara, which begins at the foot of the nation's most famous falls, is an entirely different river from the upper. The average depth is roughly three times the depth of the upper river, with some spots as deep as 150 feet. This depth, together with the added oxygenation provided by the falls and the power plants, help to attract both baitfish and predator fish.

The lower Niagara is the largest flow into Lake Ontario and as a result it serves as a magnet for every fish species in these waters. The lower river harbors massive chinook, coho, and Atlantic salmon as well as walleyes, muskies, trout, and even endangered sturgeon. Lake trout are another outstanding feature of the lower river. Working lures or baits along the bottom—jigs, emerald shiners, and artificials—can produce incredible action on lakers. Local authority Bill Hilts, Jr., Niagara County Sportsfishing Promotion Coordinator, reported a day in which he and two companions boated fifty-two lake trout ranging in size from 6 to 14 pounds in two and a half hours of drifting lures across the famous Niagara Bar area at the mouth of the river. Locating bottom structure is normally a key to catching lake trout, but big lakers are found throughout the lower river, averaging 6 to 9 pounds, with fish up to 20 pounds not at all uncommon. This is where the 39-pound 8-ounce state-record lake trout came from back in 1994.

Tip: As Bill Hilts writes, "Spoons can be very productive, with Hopkins, Kastmasters, Little Cleos, and Pirates all working. These fish aren't too selective when they're on a feed. Presentation is much more important.

Figure 7 ■ Outdoor writer Tony Zappia hoists in a fine fall-run chinook salmon taken in the lower Niagara River's Devil's Hole. *(Photo: Niagara County Tourism.)*

When fishing does get tough, though, try tipping your baits with cut bait. Leftover smelt from the freezer makes a great enticement for the fish, and new freeze-dried baits work well for taking lake trout."

To a greater degree than most other fishing locations, the lower Niagara River is a great 12-months-a-year fishery. During the traditional fall spawning run, the lower Niagara is one of the finest salmon fishing locations in the country. (Devil's Hole is a special hot spot for chinooks, with favored coho spots around the mouth of the river.) Salmon also can be caught from shore off Whirlpool State Park or the state-of-the-art fishing platform at the New York Power Authority (NYPA). Drifting lures early in the run is the first method that works, but once the runs get heavy, clumps of treated egg skein bounced off the bottom from three-way rigs is the way to go.

December through March is prime time for steelhead/rainbow trout in the lower river. Drifting eggs or egg imitations is the most popular approach, but Kwikfish and emerald minnows will also produce great results all the way down from the Devil's Hole to Fort Niagara. In both spring and fall, steelheads and brown trout are taken in large sizes and numbers in the Niagara.

White bass are a special bonus feature of the lower river in summer, with Artpark, Whirlpool, and Devil's Hole especially productive spots. Walleyes are back in the lower river in summer as well, and massive carp swirl through the shallows, waiting to challenge angler and tackle alike.

The lower river affords large areas of calm waters, which are fishable regardless of wind conditions, and consequently is a natural refuge for Lake Ontario boaters when the weather kicks up. Whitewater fishing at Whirlpool State Park and Devil's Hole State Park within the awesome Niagara Gorge offers unique and exciting adventure (although exceptional safety precautions are an absolute necessity).

Stocking programs in the lower Niagara are scientifically calculated to keep this fishery great. In 1998, for example, the DEC stocked 85,440 steelhead, 185,510 chinook salmon, 35,000 cohos, 4,000 brown trout, and

Figure 8 ■ Captain Frank Campbell took this 12-pound steelhead in the lower Niagara River, among the finest steelhead waters in the country.
(Photo: Niagara County Tourism.)

16,100 walleyes in the lower river. This generous stocking was augmented with another 90,000 walleye placed in the river by Niagara River Anglers.

Shoreline fishing access is excellent at Artpark in Lewiston, at the Lewiston Sand Dock and Fort Niagara State Park, at Joseph Davis State Park, at Youngstown Harbor and literally hundreds of other spots.

23 Hudson River, Upper

Directions: Follow I-87 to Route 4, which closely follows long stretches of the upper river in Washington and Rensselaer Counties. Route 28 leads to relatively accessible stretches of the river in Warren County.

The mighty Hudson River originates at Lake Tear of the Clouds, the highest point in New York's Adirondack Mountains, and steadily gains force and character as it flows 315 miles to New York Harbor. On its journey to the sea, the Hudson is many rivers; it is, by turns, a mountain trout stream, a bass and pike river, a canal, an estuary, and a seaport.

Long stretches of the upper Hudson flow through country so wild and so beautiful that anglers willing to hike through rough terrain are treated to vistas of how the world looked before axes and chain saws and the encroachments of civilization began to change its face. The small stream that is the beginning of the Hudson is swollen by many tributary flows, the largest and most prominent to include the Boreas, Opalescent, and Indian Rivers. Each of these is a fine trout stream in its own right, and the cold, clean water they bring to the Hudson is dappled with flashing fish.

Of the 161 miles traversed by the Hudson between its headwaters and the river's confluence with the Mohawk, more than half still lie in wilderness—remote, roadless, and seldom seen. In places, mountains line both banks of the river and the current runs swift and deep. (At the bend between West Point and Constitution Island on the lower river, the bottom is said to be more than 200 feet down.)

From the fabled Blue Ledges (below the Indian River juncture) downstream to North River, anglers enjoy rewarding trout and northern pike fishing. However, this reach of the river is primarily notable as a first-rate smallmouth bass stream.

The upper Hudson is often a solitary place to fish, with no angler visible up or downriver from the spot one chooses to cast. In many ways, the upper river is perfect for the fly or spin caster to wade, since it provides every condition conducive to fish, from pocket water to pools, riffles to flats, undercut banks to rocky structure. As the angler proceeds downstream, he or she begins to encounter the occasional deeper pool that spells smallmouth. Indeed, the fishery for bronzebacks in the upper river is superb, with scrappy fish in the 2- to 4-pound class especially susceptible to helgramites, leeches, and small crawfish worked along drop-offs and rocky ledges.

As the angler works southward, the purity of the upper river trout and smallmouth bass fishery begins to yield to a broader population. Casting artificials simulating fallfish, anglers pick up the occasional pickerel, northern pike, tiger muskies, or largemouth bass as the river flows toward Troy. Crappie are found in areas with sufficient cover and small forage fish, and significant numbers of catfish, eels, and carp—often quite large in size—are regularly taken on natural baits. (It is important to remember that from Hudson Falls downstream to the Troy Dam, all fishing is catch and release only, due to PCB contamination.)

Tip: Where you find smallmouths, mark the spot carefully in your memory and come back to it another day. Unlike lake-dwelling smallmouths, those in rivers rarely school up. Instead, an individual fish tends to pick out a quiet spot behind a rock or log where it lives, perhaps with one or two other bass. When one is caught, another moves in to take its place. As a result, prime spots tend to continue to hold bass, despite heavy fishing pressure.

The true line of demarcation between the "two Hudson Rivers" is the Federal Dam at Troy, because from Troy south to the sea, the Hudson is an estuary, with saltwater tides and an unimaginable potpourri of fish life running many miles upriver on every high tide. However, the upper Hudson—that majestic stretch from Lake Tear of the Clouds to the dam

Figure 9 ■ Sixth-grader Danny Lane is all smiles and thumbs up after catching this 11-pound 7-ounce channel catfish in the upper Hudson River.
(Photo: Daniel J. Lane, Sr.)

at Troy—is perhaps the most memorable, if not for the largest fish, then for the wilderness experience that lives long in the angler's heart.

24 Susquehanna River

Directions: The Susquehanna begins at the outflow of Otsego Lake in Otsego County. Take Route 28 west from Cooperstown along the upper reaches of the river. Follow both sides of the river's last 30 miles in New York State (in Tioga County) along Route 17 and Route 17c.

The Susquehanna is often referred to as the "Mighty Susquehanna River" along its course in Pennsylvania. Certainly no river in the northeast has more dramatically demonstrated its power for sheer destructiveness than this one, especially during the great floods in the 1930s and 1970s, when it literally inundated entire cities the size of Wilkes-Barre and Johnstown. (We clearly remember the problems Coast Guard craft experienced when their propeller blades tangled in telephone lines on Main Street in Wilkes-Barre in the flood of 1972.) However, under most circumstances, the Susquehanna is peaceful and quite beautiful as its wends it way from New York State through Pennsylvania and Maryland into the Chesapeake Bay.

The Susquehanna is, at 444 miles, the longest river on the coast of the eastern United States, and quite possibly the longest non-navigable river in the world. It begins peacefully enough in Otsego County, where several fine trout streams, including Schenevus Creek, Red Creek, Oaks Creek, and the outflow of Otsego Lake come together to form the river's quiet headwaters. In its quiet, tree-shaded beginnings, the Susquehanna is an ideal trout stream. Soon, the river begins to change its personality, growing in volume and power and becoming an ideal habitat for walleye and bass.

The river flows through portions of Otsego, Delaware, Broome, and Tioga Counties, flowing into Pennsylvania and then looping back into New York near Binghamton before recrossing the state line into Pennsylvania again. In Broome County, particular hot spots for walleye are found in Rock

Bottom and Goudy Station Dam. In neighboring Tioga County, the Susquehanna flows for nearly 30 scenic miles, with long, shallow stretches alternating with deep pools where big bronzebacks and walleyes lurk. Hot spots for both game fish include deeper water around Hiawatha Island, Nichols, Barton, and the area just east of the Pennsylvania border near Waverly. Bailey's Eddy, near the village of Nichols, is an especially productive spot for big channel catfish as well as walleye and bass. Trophy muskellunge occasionally are taken in the deep pools near Oswego.

Tip: If you're that patient angler willing to put in the time and work required to score with muskies, the Susquehanna may be just the river. Pick your spots carefully. Ideal sites in summer and fall are areas with swift current, including the upstream edges of shoals. Bars and eddies below dams can be excellent, too, since big muskies seem to find small pockets of slack water within the fast current. Fish the current edges in the river the way you might work weedbed edges in a lake. And remember to use lures in the river that work for you in lakes, including silver jerkbaits and spinner baits, especially if you are fishing at night.

Many anglers fish the river from boats, usually anchoring above favorite pools or eddies to work minnows or worms in the current. Trolling deep-pool areas with floating plugs like Rapalas and casting to structure with Mr. Twister jigs, Mepps spinners, and live bait also produce plenty of bass and walleye. When walleye are reluctant to take fancier offerings, it is good to remember the words of no less an authority on fly and plug fishing than Lee Wulff, who noted, "Walleyes are moody fish, and when the angler knows there are fish in the water he's covering but they won't strike his artificial lures, it's a good idea to resort to the worm. Worms dangling enticingly upon a hook preceded by a spinner are perhaps the walleye's greatest love."

Boating access to the Susquehanna is via ten hand-launching sites located respectively at the villages of Nichols (East), Colliersville, Sidney, Unandilla, Wells Bridge-Otego, Oneonta, Emmons, Crumhorn, Nichols (West), and Barton. Limited parking (ranging from seven to forty cars) is available at all sites.

25 Delaware River, Main Stem

Directions: Take I-84 to Port Jervis, then take Route 9 upriver for most of the length of the main stem.

The noted fishing writer Ed Zern once wrote that if forced to choose one body of fresh water to fish for the rest of his life, it would be the Delaware River. Noting that he had fished the finest rivers, lakes and streams in countries throughout the world, Mr. Zern said that none equaled the Delaware in its extraordinary variety of fishing opportunities, from its sparkling fresh headwaters high in New York State to the salt waters of Delaware Bay.

He gets no argument from the authors, who have spent literally thousands of hours fishing the lower river with great success. So many different conditions occur along the hundreds of miles of this great river's course that the Delaware could be described as dozens of different fishing venues, beginning with the sparkling streams at the headwaters of the East and West Branches of the river, the massive reservoirs that change the contours of each, and of course, the main stem of the river, to which this section is addressed.

The main stem of the river is formed by the junction of the East and West Branches of the Delaware near the village of Hancock, and flows more than 330 miles from this birthplace to the sea.

The upper section, comprising 80 miles of wild, beautiful river, is especially known for trout fishing, with trophy rainbows and browns flourishing in the cold, clean water released downriver from the Pepacton and Cannonsville Reservoirs. Indeed, as noted by DEC authority Bob Brandt, the main-stem Delaware above Kellham's Bridge is a world-class trout fishery.

However, it is the shad run that excites the most angling activity, with thousands of anglers working shad darts and gaudy streamer flies in likely spots throughout the river, all the way up the main stem and the East Branch to Pepacton Reservoir. The best fishing occurs in late April and continues through the month of June. The shad run lasts longer in the upper

Delaware than the lower sections of the river because of the colder water and provides an often spectacular day on the water. More than incidentally, added bonuses often spice the day when large trout and walleyes accidentally take shad darts.

Tip: Bundle up and fish the upper river for walleyes when the water is cold in fall and spring. Hardy anglers willing to endure difficult climatic conditions score well on "eyes." The axiom here is not unlike the familiar admonition to duck-hunters: The worse the weather, the better the hunt. Walleyes tend to be caught in cold, blustery weather. (Remember that the walleye season closes from March 15 until the first Saturday in May.)

During warmer weather, a few walleyes are taken in cooler, deeper pools of the river, most often on live bait, including leeches, minnows, and night crawlers.

Muskellunge also are taken in the upper reaches of the river, most often on heavy gear and a wide range of artificials. However, the muskie angler who ignores live suckers and exceptionally large fallfish may be missing a good bet.

In warmer months, when shad and walleye fishing diminishes, more angling attention is focused on the river's smallmouth bass. Like trout, these aggressive, hard-fighting fish have a wide range of appetites. Small crawfish, hellgrammites, and fallfish (baby "chub") are at the top of the menu; however, they also are taken on night crawlers and a wide variety of artificials, of which small Rapalas and silver Mepps are our personal favorites—especially in rocky sections of the river.

Another warm-water favorite is the common (actually, quite uncommon) carp. These great fish, which grow to mammoth proportions (the New York State record is 50 pounds 4 ounces) remain the best-kept secret in all of American angling. Pound for pound, river carp—and in our experience, especially Delaware River carp—are among the most powerful freshwater fish that swim. They are also among the most obliging: feeding most actively in hot weather when most other fish are virtually dormant, taking a wide range of bottom-fished baits, and fighting in long, tearing runs and tackle-threatening lunges. The number two game fish (after salmon) in England, the carp suffers from a poor image in the United States, not because it lacks strength, but because that strength is so little known.

Access to the river is excellent both for boat and shore fishing. Boat and canoe rentals are available at many sites along the river. Boat launch facilities maintained by the DEC include sites at Calicoon off Route 97, at Narrowsburg in Sullivan County off Route 97, at Skinners Falls off Route 97, 6.5 miles north of Narrowsburg, and at Sparrowbush in Orange County, northeast of Port Jervis, off Route 97.

26 Hudson River, Lower

Directions: Routes 9 and 9W parallel virtually the entire length of the lower Hudson, often in full sight of the river, from Manhattan upriver to Troy.

The Hudson River most people know is the 154-mile stretch from Troy down to Manhattan. Here the "lower Hudson" is essentially an arm of the sea—an estuary in which ocean tides run all the way to the federal lock and dam at Troy with virtually no drop in surface elevations between the two points.

The biological productivity of this estuary is staggering. Marine and freshwater species feed side by side in populations so massive as to comprise the greatest single wildlife resource in New York State. Nearly 200 species of fish have been identified throughout the estuary. Striped bass, bluefish, largemouth and smallmouth bass, pickerel, pike, yellow and white perch, tommycod, crappie, catfish, eels, shad, and carp are just the fish caught most frequently by Hudson River anglers. Each year, local anglers also take weakfish, blackfish, fluke, flounder, jack crevalle, and tiger muskies. And even this list only begins to hint at the bounty of the lower Hudson.

Striped bass are the largest fish commonly caught in the Hudson, as evidenced by catches in two consecutive Southern New York Fishing Derbies by an angler named Walter Chudkosky. Mr. Chudkosky took first prize honors in 1995 with a 46-pound 11-ounce monster just south of the Newburgh Beacon Bridge; then took first prize again in 1996 with a

Figure 10 ■ Robert W. Spanburgh, Sr., caught this New York State Inland record (48 pounds 3 ounces) striped bass in the Hudson River on a 14-inch live herring and a heavy-duty Carolina rig. *(Photo: Robert W. Spanburgh, Sr.)*

43-pound 15-ounce bass. (The 1995 fish set the New York State record, which stood until Robert Spanburgh, Sr. took a 48-pound 3-ounce striper—also in the Hudson—on a live herring in May of 1998.)

Stripers are found throughout the estuary. The season opens in early March but most fish taken in the colder water of early spring are short. However, when the weather warms in late April and May, thousands of big females between 10 and 40 pounds are taken. Significantly, the striper population in the Hudson is thriving. In peak years, upwards of 800,000 stripers power up the Hudson to spawn. These big females respond to lures—especially Rebels and Rapalas—out of annoyance rather than hunger. Live and cut bunker, bloodworms, sandworms, and eels are top-producing baits.

At about the same time the monster stripers are being taken, parallel excitement accrues to the shad run. The Hudson River stock is considered the largest run on the Atlantic coast as upwards of a million sea-run fish come back to the river to spawn each year between Kingston and Albany. Along the way, thousands of anglers work shad darts and shad flies along channel edges and over flats as slack tides approach. The shad move upriver along the edges of flats, shoals, and sandbars as the current slows approaching high tide. They return to deeper water as the tide ebbs.

The returning shad average about 4 pounds; however, fish of 6 pounds are not uncommon. Not surprisingly, the New York State record shad was caught in the Hudson; a fine 8-pound 14-ounce specimen taken by Andrew F. Sheffer in late April of 1989.

Tip: Tidal movement is the critical factor is catching Hudson River shad. The best time to fish is the slow-current period just before the change in direction of tidal flow near the high and low slack tide. Normally the action builds for about an hour as the current slows. Best fishing often comes during the hour of dead slack and tapers off as the current builds again.

The other avidly sought game fish in the Hudson is black bass—both largemouths and somewhat less populous smallmouths. Significantly, the superb black bass fishing was essentially a well-kept secret until bass fishing tournaments in 1984 started producing dramatic results.

The salinity in the lower estuary is generally too high for black bass. However, the roughly 75-mile stretch from the Troy dam down to Newburgh, prominently to include the tidal portions of such fish-rich tributaries as Esopus Creek, Fishkill Creek, and Rondout Creek, provides bass habitat that is both extensive and diverse. Abundant weedbeds are found along the shores of this stretch of the river, and where weeds abut hard structure such as rocky points or old bridge pilings, largemouth bass are likely to be found. As in other waters, the Hudson River smallmouths feed and shelter around rocks. Artificial rock piles and the bases of beacon towers are prime bronzeback hangouts.

Unlike more typical bass waters, the Hudson's black bass fishing is entirely sensitive to tides. Roland Martin, the reigning impresario of television bass fishing, called fishing the tide changes, "definitely the key to catching a lot of bass on the Hudson." As usual, Mr. Martin knows whereof he speaks, having won the Hudson River B.A.S.S. Invitational some years back.

The Hudson also features wonderful white catfish and carp fishing, from its most northern freshwater regions all the way down to the George Washington Bridge. Panfish, prominently to include white and yellow perch, are taken in great numbers.

Boat launch ramps are found at Corning Preserve and Coeymans Landing in Albany County; Coxsackie, Athens, and Catskill in Greene County; Hudson, North Germantown, and Germantown in Columbia County; Saugerties (4 facilities), town of Ulster, Connelly/Rondout Creek (2 facilities), and Marlboro in Ulster County; Rhinebeck, Staatsburg, Andros River Road, Poughkeepsie, New Hamburgh, Chelsea, and Beacon in Dutchess County; Newburgh (2 facilities) and Cornwall in Orange County; Stony Point (2 facilities), West Havestraw, Nyack (2 facilities), Piermont (2 facilities and a superb location for pier and dock anglers) in Rockland County; and Peekskill, Lents Cove, Montrose, Croton-on-Hudson,

Ossining, and Tarrytown (2 facilities) in Westchester County. A special wheelchair-accessible angling facility is maintained by the DEC at Germantown in Columbia County.

All in all, the lower Hudson is almost unexcelled as a fishing venue. In size and in diversity of sportfish species, it is rivaled only by the Great Lakes and the major salmon rivers to the north. In accessibility to millions of New York State anglers, it is rivaled by none. That anglers can flycast for brook trout at the river's pristine headwaters, compete in national black bass tournaments in its mid-point, and catch record-threatening sea-run stripers at its mouth—the latter virtually in the shadow of skyscrapers—tells us everything we need to know about this great river.

Anglers should be aware of specific fish-consumption advisories, which warn against eating any fish taken from Hudson Falls to Catskill (except American shad), no more than one meal per month of most gamefish species (walleye, striped bass, lagemouth and smallmouth bass, bluefish, carp, white perch, white catfish, etc.) taken from Catskill south to the George Washington Bridge, and no more than one meal per week of other species.

Smaller Rivers

"An undisturbed river is as perfect as we will ever know, every refractive slide of cold water a glimpse of eternity," Thomas McGuane wrote in *Midstream*. We see this perfection especially in the smaller rivers of New York State. For us as for many anglers, rivers and streams are constant sources of fascination, challenge, and excitement. With smaller rivers, a particular deep channel, a long stretch of rapids, an undercut bank, a wide, quiet pool, the memory of a fine fish taken on an equally fine day live in our memories and call to us to return.

27 Chateaugay and Salmon Rivers (Franklin County)

Directions: In Franklin County, take Route 30 to the Salmon River at Malone; follow Duane Road along the river north of Malone, and River Road south. For the Chateaugay, turn right at Malone onto Route 11, which crosses the river.

Discussions about New York State trout streams tend to spotlight the Beaverkill, the Esopus, the Willowemoc, and the West Branches of the Delaware and Ausable Rivers. However, there are two small rivers in Franklin County that rival any of these better-known waters, not only in numbers of browns, rainbows, and brookies, but in size as well.

The Salmon River provides about 20 miles of prime trout water before it empties into the St. Lawrence; the Chateaugay, found in the very northeastern corner of the county, is a few miles shorter, but no less productive. Aside from proximity, both rivers share key characteristics. Water sources for both are springs in the foothills of the Adirondacks; consequently the water is cool and clean, with stable flows all year round. Both are scientifically managed by the DEC and regularly stocked. (In a recent stocking season, the DEC placed just under 20,000 brown, rainbow, and brook trout in the Chateaugay, all "keeper-sized" from 8.5 to 14.5 inches. The Salmon River received just under 34,000 trout ranging from 8.5 to 15 inches.) Both have ample food resources for sound trout growth, including a dependable supply of emerging insects each spring and summer. Both feature falls, deep pools, pocket waters, and long stretches of riffles to warm the heart and stir the imagination of any trout angler.

It should be noted that the Salmon River referenced here is the "other Salmon River," not to be confused with the larger Salmon River in Oswego County, which flows into Lake Ontario (and is treated in a separate section of this book). However, for serious trout anglers—especially devotees of the dry fly—this may be the Salmon River of choice.

The DEC's Region 5 Aquatic Biologist Richard Preall attests to the size and populations of trout, with plenty of 5-pound browns in both rivers plus excellent populations of rainbows and brookies. "The brook trout are basically wild fish from the tributary streams, ranging up to 15 inches," he notes. "In our recent survey of the Salmon, we found plenty of browns up to 19 inches, and the Chateaugay is home to plenty of fine wild brown trout." Preall adds that a number of tributary streams that feed or lie between the two rivers all along the Canadian border are themselves excellent trout streams.

Tip: Presenting fly, spinner, or bait in the right place is a critical key to taking river trout. Study the flow for small eddies, cuts, or slough edges where minnows and other forage foods most likely become available to larger trout. Then present your offering naturally, to tempt fish out of likely lies or holding stations under cutbanks and around structure.

While the Salmon, like the Chateaugay, is notably a trout river, it also provides some opportunity for warm-water species—from Malone north to the St. Lawerence. Two impoundments in the Malone area shelter good populations of northern pike and tiger muskellunge. The last several miles of the river also provide decent fishing for pike and tiger muskies, courtesy of stocking programs by the DEC. Between the dam at Fort Covington near the Canadian Border to the St. Lawrence River also offers better-than-average smallmouth bass action.

Boat access on both rivers is largely irrelevant, since their size and structure primarily indicate wading or fishing from shore. The two impoundments near Malone are large enough to be worked from small boats. Access in general is first rate with long stretches of both rivers posted for public use.

The unspoiled beauty and wildness of these two rivers recalls lines from R. Palmer Baker's little book, *The Sweet of the Year.* "In the recollection of the trout fisherman," he wrote, "it is always spring. The blackbird sings of a May morning. The little brook trout jump in the riffles, and the German Brown comes surely to the fly on the evening rise."

28 Oswegatchie River

Directions: N.Y. Route 3 crosses the confluence point of the river with Cranberry Lake and touches the river again at Fine. Routes 11 and 812 follow the river from Richville to Ogdensburg.

The headwater ponds that give birth to the Oswegatchie River are so remote and irresistible that anglers trek through miles of forest trails—burdened with fishing gear and lightweight canoes—in search of their pristine waters and wily trout. The largest of these, 30-acre Darning Needle Pond, requires a hike of just over 2.5 miles; the next largest, Cowhorn Pond, a hike of 6.5 miles. Yet anglers are more than willing to swap perspiration and fatigue for the true wilderness adventure that beckons at the end of the trail.

The Oswegatchie traces a sinuous path of more than 100 miles from these ponds to and through Cranberry Lake to its ultimate appointment with the St. Lawrence River. Along the way, its personality changes again and again, alternately wide marsh and narrow corridor, calm water and swift, shallow and deep, open and rocky. No fewer than seventeen dams, ranging from ruined old structures to modern hydroelectric dams, alter the river's size and flow, often in dramatic ways.

Whereas the most remote fishing on the Oswegatchie may be in its source ponds, the best fishing is often found in considerably more accessible stretches. At the Newton Falls Impoundment, for example, the angler finds a pondlike setting teeming with smallmouth and largemouth bass and northern pike. In the stretch between Fine and South Edwards, fast water and rapids keep canoeists off the river but provide fine habitat for brook trout, brown trout, and smallmouth bass, together with swarms of yellow perch and other panfish. Along the way, anglers who know the river fish the "spring holes"—often the local reference to confluences with tributary streams—especially at Carter's Landing, Wool Creek, Cage Lake, High Rock Creek, Dorset Creek, and Otter Creek.

Tip: The water in these spring holes remains cooler as spring turns to summer; thus these spots are inevitably attractive to trout. In early spring, live bait—especially small night crawlers and minnows—are extremely productive when worked along the edges and drop-offs. As summer progresses, small spinners—particularly silver and bronze Mepps—and other artificials may be more productive.

At the confluence of the Oswegatchie and Robinson Rivers, approximately 3 miles above High Falls, the fishing remains excellent right through spring and summer as the Robinson rushes cool water and forage fish into the Oswegatchie.

The most significant feature along the Oswegatchie is Cranberry Lake, a lake of more than 11 square miles that is entered by the river about 25 miles downstream from the river's source. Cranberry Lake is itself a fine fishing venue, with the trout populations regularly augmented not only by the inflow from the Oswegatchie, but also regular DEC stockings of tens of thousands of brook and brown trout. (Brook trout also are stocked in the West Branch of the Oswegatchie at Crogan and Diana, and brookies and browns in the main body of the river at Clifton and Fine.)

Perhaps the best fishing of all is found in the 5-mile stretch from Eel Weir State Park to the Ogdensburg, just above the river's juncture with the St. Lawrence. Here powerful muskellunge prowl the depths, swimming cheek by jowl with outsized northern pike, walleye, smallmouths, crappie, and yellow perch.

The Oswegatchie and Cranberry Lake are modern parables about rivers and lakes, both humankind's power to affect them and nature's ability to persevere. In the early part of this century, one of the most cherished wilderness adventures was a canoe trek up the fabled Oswegatchie Inlet of Cranberry Lake in quest of lunker brook trout. However, a new dam was built that doubled the lake's size and drowned the inlets. Worse still, in the mid-1940s, yellow perch were inadvertently introduced into the river, and the voracious habits of these competing fish almost immediately eroded the brook trout fishery, especially in the lake and its inlet river section. In a curious way, however, an environmental accident made the lake a more favorable habitat for brookies. Acid rain in the late 1970s drastically reduced the yellow perch stocks. As yellow perch declined, the brook trout population rebounded (aided and abetted by DEC stockings). In the 1960s, smallmouth bass were stocked as well and a thriving "bronzeback" fishery was established, such that smallmouths now are found throughout the 100-mile length of the river.

One of many attractive qualities of the Oswegatchie is its exceptional accessibility. Both developed (boat ramps, parking, etc.) and undeveloped access points are maintained at strategic points along the river, with especially heavy concentrations in the stretch from Newton Falls to Edwards (above Cranberry Lake) and DeKalb to Ogdensburg at the river's terminus. What is glorious about the river is that the fishing is excellent along most of its length, especially in some of its most easily accessed points.

29 Salmon River (Oswego County)

Directions: Take Route 81 to exit 36 North to Route 13, which traces the river-bank from Altmar all the way to Lake Ontario.

The Salmon River is most aptly named. World-record catches in this magnificent fishery include the all-tackle and 30-pound-line-class coho salmon, a 33-pound 4-ounce monster taken in 1989, and the state record chinook, a 47-pound 13-ounce salmon caught two years later. Certainly these record fish are the stuff anglers' dreams are made of. However, what is infinitely more exciting is that extremely large fish are not unusual. Indeed, chinooks in the 30-pound range are commonly caught during the spawning runs and occasionally a fish that tops 40 pounds will be taken.

The watershed of the river is located on the western slope of Tug Hill, primarily in Lewis and Oswego Counties. The river flows easterly for 44 miles from its headwaters in northern Osceola through Pulaski to its rendezvous with Lake Ontario. The state has acquired more than 20 miles of public fishing rights on the upper Salmon and its tributaries, including the North Branch of Salmon River, the lower Mad River, and Prince, Mallory, Stony, and Fall Brooks. However, it is the river's final 27 miles from the upper reservoir to the big lake that provides the great salmon fishery. Each year, thousands of anglers flock to the river in September and October for the salmon run, taking fish on flies, spinners, egg sacs, and a wide variety of floating and diving plugs.

One skilled angler, Bill Zarola of Wilkes-Barre, Pennsylvania, recently was pictured in his local newspaper holding a salmon in the 35-pound range—a fish he returned to the water to fight again. So sensational is the fishing here, Zarola said, that he and his two brothers hooked fifty salmon on heavy-duty flyrods during one recent trip.

Tip: The river bottom is slippery year-round. Felt soles on the bottom of your waders won't work. You need studded rubber grips—"corkers"—that slip on over your wading shoes for better purchase when wading. And carry a wading staff to improve footing and prevent nasty falls when you do slip.

Figure 11 ■ Braving bitterly cold weather, Karl Elsner, Jr., caught this 16-pound steelhead on a 4-pound test in the Schoolhouse Pool of the Salmon River.
(Photo: Karl Elsner, Jr.)

While the height of the salmon season lasts from mid-September to mid-October, the Salmon River is truly a great year-round sportsfishing venue. Certainly it is best known for its chinook and coho salmon. However, the river actually boasts three additional major sports fish as well: steelhead, Atlantic salmon, and brown trout.

Two strains of steelhead ply the river, ranging in size from 5 to 25 pounds. The first strain, Washington (winter run) steelhead, begin entering the river from Lake Ontario about the time the coho and chinook fishing starts to wane, roughly mid-October. The run continues right through the winter into March, with spawning occurring in early April. Skamania (summer run) steelhead begin entering the river the following month, and remain to spawn the following spring. October through mid-May is the hottest time to fish for winter-run steelhead, whereas summer-run steelhead fishing is superb from late June through April. It was here that Manny waged a titanic battle with the largest steelhead he ever hooked. After a long struggle, the hook finally pulled free, leaving an angler shaking his head in admiration and a great fish free to fight another day.

Atlantic salmon, which run up to 20 pounds, are fall spawners and provide great sport for Salmon River anglers in mid-summer. Brown trout, which also spawn in the fall, enter the river from mid-September through mid-November. These fish range in size from several pounds to more than 20 and provide wonderful fishing, especially in "fly-fishing only" areas above the county Route 52 bridge in Altmar to Beaverdam Brook and the upper section adjacent to county Route 22 above the hatchery down to the Lighthouse Hill Reservoir tailrace. The river also is home to spectacular rainbow trout. In fact, a 16-pound 8-ounce rainbow was caught in the river by a fly-fishing angler using a 2-pound test tippet—another world-record catch.

Finally, good smallmouth bass fishing can be had throughout most of the year in stretches of the river below Pulaski, with the best catches usually recorded from the estuary at the mouth of the river. In the weedier backwater areas of the river, largemouth bass and northern pike complete the action in this extraordinary fishery.

The DEC works assiduously to assure that this great river continues as one of the premier salmon fishing venues in America. In one recent stocking season, for example, the DEC stocked more than 300,000 chinook salmon between 3.5 and 4 inches long in the river at Richland County alone.

The river is broadly accessible, with access points on both the north and south banks of the river managed by the DEC. River-access parking areas can be found on the south bank in Pulaski and Altmar, and at the Trestle Pool South and Sportsmen's Pool South.

30 Mohawk River, Lower

Directions: Follow N.Y. Route 5 along the northern bank and Route 5S along the southern Bank of the Mohawk from the Five Mile Dam downriver to Lock 8. Take U.S. Route 87 across the river between the Crescent Dam and Lock 7.

For the angler thirsting for wild vistas and massive fish, the Salmon, the Hudson, or the Delaware River is just what the doctor ordered. But for those seeking a wonderful, diverse fishery in more peaceful settings, with unparalleled access both for shore fishing and boat launching, the lower Mohawk River is definitely the place to be.

The reasons begin with the exceptional array of species awaiting fly, plug, or bait. The most numerous and most popular are smallmouth bass, which are abundant virtually throughout the river's 76-mile flow from the Crescent Dam to the river's confluence with the Hudson River. Bronzebacks average 11 to 13 inches, but the occasional 4-pounder rips line off reels and sends hearts pounding. In recent studies, catch rates on the Mohawk roughly doubled those of the fabled St. Lawrence River, and the numbers keep getting better every year. Walleye and yellow perch are no less common, being found in every portion of the river and most especially in the cooler, faster water below the dams. Anglers also take largemouth bass, northern pike, tiger muskie, white perch, carp, suckers, and several varieties of panfish in various stretches of the river, all in significant numbers.

The lower Mohawk is significantly different from its sister rivers in that it is subdivided and controlled, with four dams and eleven locks built along its 76-mile course. Moving from one section to another requires "locking" through a navigational lock, a simple operation in which you motor into the lock, stay in place while the water level is raised or lowered, and then motor out of the lock. Fishing boats of virtually any size are allowed on the river, and as discussed a bit later, access, parking, and boat-launching sites are unmatched anywhere else in the state. Significantly, miles of river are encountered between locks such that enormous pools are created for fishing pleasure. Even though the river flows by major urban centers such as Albany and Schenectady, the water quality is excellent.

Shore fishing is especially popular at the downstream sides of locks because these areas have good road access, plenty of parking, and the kind of structure that holds fish. Downstream of Lock 8 to the Crescent Dam, a distance of about 20 miles, shore fishing opportunities are most plentiful.

Tip: A few simple precautions will keep you from losing a trophy fish. First, retie hooks from your last outing, snipping off the last few feet of line, which may have been frayed or nicked. Second, when fighting a good fish, keep your rod tip up to have a margin of error if your lunker makes a sudden lunge. Third, never net a fish before it is ready.

Anglers can shape a day's fishing by selecting a particular stretch of the river. For example, in addition to smallmouth bass, largemouth bass are dominant on the lower river between Lock 7, the southernmost lock, and Crescent Dam, which blocks the river's waters a final time before it flows into the Hudson. Walleye, while found throughout the river, are more abundant upstream of Lock 8, and as always, are most susceptible to lure or bait at first light and last light. The DEC stocked tiger muskies initially in the Lock 8 to Crescent Dam area and now is extending the range of stocking into other upriver stretches, working to establish a trophy fishery. Although catches are not numerous, results are promising, with several fish over 18 pounds having been taken to date. (Trolling artificials produces occasional tigers, but our money is on live bait, worked carefully around the edges of weedlines.) The 11-mile stretch between Lock 7 and Lock 8 is especially beautiful and productive.

Tributaries to the river provide additional fishing possibilities. For example, Schoharie Creek flows into the Mohawk just above Lock 12, and the juncture provides especially exciting bass fishing throughout the season, as well as super walleye fishing in early spring.

The stretch of the river from Lock 13 up to Lock 16 is perhaps even more appealing, with plenty of game fish and light fishing pressure. This lovely stretch of the river recalls a statement from *A River Never Sleeps*, in which the author admitted, "Perhaps fishing is, for me, only an excuse to be near rivers."

Big, powerful carp are abundant here as in every section of the Mohawk and should not be overlooked as game fish of rare fighting potential. Panfish, most prominently including suckers, yellow perch, and rock bass, are also found throughout the river.

Regarding access, no fewer than a dozen public boat-launching sites are available between the Crescent Dam and Lock 16. In Saratoga County, a first-rate facility is located at the end of Flight A Road. Three public launch sites are located in Schenectady County: off Rosendale Road, at the end of Lock 7 Road; at the Northeast side of Freemans Bridge, at the end of Mohawk Avenue; and off Route 5S in Kiwanis Park. Six public launch sites are operated in Montgomery County, respectively, between Locks 8 and 10 up to Locks 15 and 16. Specific locations are: the southeast side of Lock 10, off Route 5S; off Quist Road in Amsterdam; off Route 5S at the mouth of Schoharie Creek; the south side of the Route 10 bridge; off Route 80 on road to Nelliston Plant; and at the end of Bridge Street at St. Johnsville Marina.

Moreover, no body of water in New York State offers more wheelchair-angler fishing sites, including accessible platforms and piers equipped with safety rails and wood or hard-surface walkways to the water's edge. In Herkimer County alone, wheelchair-accessible facilities are maintained on the river at Moss Island Park in Little Falls adjacent to Lock 17 of the Barge Canal, at the village of Herkimer on East Washington Street, in Ilion, in the village of Mohawk at Fulmer Creek off Route 5, and in Frankfort at the end of Fox Street. Similar facilities are maintained in

Montgomery County at Amsterdam, at Lock 10, and at Nelliston; and along the upper Mohawk in Oneida County (at Lock 20 at Canal Park) and just west of New London at Lock 22 on Wood Creek Road.

We should add, if only in passing, that the upper reaches of the Mohawk, flowing more than 90 miles down from the southwestern Adirondacks to St. Johnsville, are wonderfully scenic and considered by some to be trout waters of unparalleled richness. However, it is the lower Mohawk that serves the greatest angling populations. And certainly it is the lower river that provides the most diversity and, to us, the most exciting fishing potential. This is why we have chosen to chronicle the tamer but more productive lower river here. We commend it to our readers without hesitation.

31 Genesee River

Directions: Take Route 19 along the riverbank in Allegany County from the Pennsylvania line north to Fillmore. Follow Route 383 along the portion of the river above its junction with Lake Ontario.

The Genesee is the only river that completely traverses New York State, traveling north from its headwaters in northern Potter County, Pennsylvania, to its mouth at Ontario Beach on Lake Ontario. Genesee is an Indian word meaning "beautiful valley," and the terminology could hardly be more appropriate. The river courses through broadly varying terrain, from a gentle river valley to a gorge so deep it is called the "Grand Canyon of New York," and most of it remains to this day beautiful indeed, and broadly inviting to anglers bent on fine bass and trout fishing.

The headwaters of the river comprise three tributaries, which join near the town of Genesee, just south of the New York border. The river then flows across western New York on its way to and beyond Rochester, where it enters Lake Ontario. Along the way, the river benefits from deposits not only of cold, clean water but also native and stocked browns and rainbows from such outstanding trout waters as East Koy and Wiscoy Creeks. Indeed, the Genesee River system has been called the "premier drainage in western New York for the fly-fisher" by no less an authority than author/angler Charles R. Meck.

The most prominent feature along the Genesee is Letchworth Gorge, a 21.5-mile canyon with almost vertical 300- to 400-foot rock walls—the Mt. Morris highbanks—and three falls, which measure 71, 107, and 70 feet, respectively. So beautiful is this area that New York established Letchworth State Park to preserve its pristine state. Of interest here, the smallmouth bass fishing is superb in the stretch of the river above the state park in Livingston County. Below Mt. Morris Dam, fine populations of walleye rival the bronzebacks for angler attention.

The river is heavily stocked by the DEC. In Allegany County, for example, a recent stocking season saw more than 32,000 brown and rainbow trout from 8 to 14.5 inches stocked at Amity, Wellsville, and Willing. Downriver, the stocking is much more diverse. For example, in Monroe County, 26,000 steelhead, 153,000 chinook salmon, and 22,000 coho salmon were stocked near Rochester.

The fishing is first rate along the river's entire length. Trout anglers score well in the river's heavily stocked upper reaches, between the Pennsylvania line and the Belmont Dam. The upper river has fine insect hatches of Hendricksons, blue quills, caddis, sulphurs, slate drakes, light cahills, tricos, and olive duns, among others. The upper river features many long runs and deep pools that harbor good populations of trout. A 2.5-mile catch-and-release section a few miles downriver from the Pennsylvania border is especially productive for fly anglers.

Tip: A thoughtful study of a particular flow and a natural presentation of artificial or natural bait is important in such sections of the Genesee and other trout streams. As outdoor writer Jerry Gibbs points out, "Like all living creatures, trout learn when and where regularly occurring food items appear in their environment. Those patterns affect where the fish locate seasonally and also where they set up relative to specific ambush points. They dictate areas a trout will regularly patrol in search of food. Though there will always be some fish that will eat a bait presented unnaturally or at the wrong time, you'll do far better imitating nature with your presentations."

As noted, the mid-regions of the river, from the Letchworth State Park north to the hamlet of Genesee, produce fine smallmouth bass and walleye fishing. And where the river enters Lake Ontario, there is superb steelhead fishing. In addition, anglers catch brown trout, lake trout, and even coho salmon right from the two large piers at the river's mouth and from shore.

Access to the upper Genesee is first rate. As pointed out in the directions, the stream is paralleled for many miles by a paved road (Route 19). Moreover, many convenient parking areas and bridge crossings are found along its course. Among many boat launches are two small sites in Monroe County operated by state parks. The first is situated on county Route 252, 2 miles east of Scottsville; the second on Route 251, 2 miles southeast of Scottsville. Both are hand-launch sites with limited parking.

32 Cattaraugus Creek

Directions: Take U.S. Route 20 across Cattaraugus Creek at Irving, just before it enters Lake Erie in Sunset Bay. Turn onto N.Y. Route 438, which parallels the creek.

Cattaraugus Creek is generally defined in angling superlatives. It is one of the most fish-rich of all tributaries to Lake Erie. It is said to be one of the two best coho fishing tributaries of New York's Great Lakes (the other being the Salmon River). It has the largest annual chinook salmon runs out of Lake Erie. And in its upper reaches, it is a truly great trout stream.

The Cattaraugus flows through four New York counties, picking up volume, stocked fish, and cold, oxygenated water in each of them. A significant portion of the main stem of the creek flows through the Cattaraugus Indian Reservation. (This is significant because stream anglers require Seneca Nation fishing licenses if they want to fish on reservation lands.)

Steelhead fishing is best in spring, with a smaller but still sensational run in the fall. Steelhead are generally thought to be sea-run rainbow trout; that is, rainbows that spend their adult lives at sea and then reenter streams to spawn. However, a huge number of steelhead spend their lives not in

Figure 12 ■ Greg Lowry caught and released this fine 24.25-inch rainbow trout after another fish of similar size broke his line. *(Photo: Greg Lowry.)*

salt water but in the immense expanses of the Great Lakes. The fish that come powering up Cattaraugus Creek each spring and fall are this variety, and they are often large and always spoiling for a fight. So memorable is the fighting power of these great fish that they sometimes move anglers to dramatic and even poetic language. In *Excalibur: The Steelhead*, Paul O'Neil wrote, "The steelhead can hurtle into the air a split second after he is hooked, and flash hugely out in the murk, like the sword Excalibur thrust up from the depths—at once a gleaming prize and a symbol of battle."

The fighting ability of these great fish proceeds out of their speed. In fact, they are faster than any other freshwater fish, having been clocked at 26.8 feet per second. A hooked steelhead will leap repeatedly, often clearing the surface by 2 to 3 feet. The average steelhead weighs 4 to 6 pounds, but some fish taken in the Cattaraugus weigh 20 pounds or more. The bait that probably takes the most steelhead is salmon or trout egg sacs, although many are taken on flies and artificials as well.

Tip: When bait fishing the Cattaraugus, make sure your fly or spinning rod has enough backbone to absorb the shock of the steelhead's strike and surging run. Cast the bait upstream and follow it with the rod tip as it drifts downstream. When it stops, set the hook.

Coho salmon, which move to progressively deeper water offshore in Lake Erie during the summer, come powering back to spawn in the areas where they were stocked or hatched in fall. The spawning runs of sexually mature fish peak in October and early November, and anglers take fish on flashy spoons, deep-diving plugs, egg sacs, egg imitations, and artificial flies in faster upstream water areas. The spawning run hits its peak as the leaves turn to gorgeous autumn colors, and the fishing is as exciting as the environs are beautiful around Cattaraugus Creek.

We began by noting that Cattaraugus Creek is often spoken of in superlatives. Certainly this is true of the stocking efforts that the DEC expends to assure the continued success of the fishery, not only in the creek but in the great lake it feeds. Nowhere is this truer than in Chautauqua County. In one recent stocking season, the DEC stocked 500,000 3-inch chinook salmon and more than 1,600,000 walleye fry at the stocking point in Hanover in Chautauqua County. In Erie County, the Cattaraugus received another 72,000 steelhead plus 7,700 brown trout ranging in size from 8.5 to 14.5 inches. In Wyoming County, nearer the headwaters, Cattaraugus Creek received another 7,000 brown trout, and in Cattaraugus County, 1,200 larger browns were stocked. The coho stocking is especially significant, since natural reproduction, which occurs in New York waters, is too limited to support a viable fishery without assistance from stocking programs.

The DEC maintains a small boat launch for canoes and shallow draft boats only (no motors, hand-launch) north of the village of Otto on county Route 11.

As noted, access is an issue on Seneca Nation lands, since a Seneca Nation fishing license must be acquired. However, cost and inconvenience pale in comparison to the spectacular fishing on the creek, especially in spring and fall. One hot spot where public fishing is available is the 1,200-foot breakwall on the south side of Sunset Bay.

33 Delaware River, East Branch

Directions: Take N.Y. Route 30 around the Pepacton Reservoir and downstream to the village of East Branch. From this point to Hancock and the end of the East Branch, follow N.Y. Route 17.

The East Branch traces a serpentine course from its headwaters below Roxbury to a point south of Margaretville, where the river was impounded to form the Pepacton Reservoir. From the reservoir south, the river runs just over 30 miles to Hancock, where the main stem of the Delaware is born. Above the reservoir, large brown trout migrate out of the still waters upstream into the river and its tributaries each fall to spawn. Large fish often attack lures and natural baits in surprisingly small tributary streams.

The upper miles of the river closest to the reservoir, where discharges from the bottom of Pepacton chill the water, are a fine trout fishery. Where the water is coldest, fine chunky brook trout thrive. Good-sized rainbows and browns, stocked and wild, are found the length of the river. A number of fine streams contribute clean, oxygenated water and stocks of trout to the East Branch. The most noted of these is the Beaverkill, which joins the river at East Branch village.

The DEC works to keep the trout fishing excellent. In 1997, browns ranging in size from 8.5 to 14 inches were stocked in the East Branch in sixteen separate stockings in Delaware County.

Unlike the West Branch, the East Branch features fine fishing for other game fish as well, especially below the reservoir. In spring, the annual shad migration begins, with incredible numbers of 3- to 5-pound buck and roe shad powering their way up to traditional spawning grounds in the East Branch. Spawning commences at water temperatures above 54 degrees and generally ends when temperatures reach 70 degrees. The fish tend to school up in pool areas and are most often taken on shad darts, flutter spoons, or bright streamer flies, fished near the bottom.

Smallmouth bass also lurk around rocky ledges and drop-offs in the upper East Branch. Fine bronzebacks, many in the 3-pound class, are taken on a variety of artificial and natural baits, the most deadly of which is inevitably a small live crawfish.

Walleye are taken in the lower reaches of the East Branch, principally in early spring and late fall, when water temperatures dip below 45 degrees. East Branch walleyes generally run between 18 and 22 inches, but more than a few 6- or 7-pound fish are taken during the season.

Tip: Use a light, low-visibility mono leader for walleyes, especially when fishing in clear water. Many commercial rigs have leaders that are much too heavy. Walleyes are more likely to see a heavy leader and shy away from bait. A 6-pound leader is adequate in most situations, but in extremely clear water or when walleyes are sluggish, a 4-pound leader may improve your odds. If you happen to be "blessed" with clear weather during a fishing trip, try to fish at first light or the last hours until and just after dark.

Access to the East Branch is excellent, especially on the lower section, where almost the entire 13-mile stretch is open to the public via state easements and miles of privately owned but unposted riverbank.

34 Neversink River, Main Stem

Directions: Take Route 42 (Hasbrouck Road) along most of the eastern bank of the Neversink's main stem to Route 53, which crosses the river at the foot of the Neversink Reservoir in Sullivan County.

The Neversink is a storied trout stream in all of its sections, certainly to include its wilder East and West Branches. However, since so much of both branches is in private hands and inaccessible to the public, we concentrate here on the main stem of the river, commencing at the foot of the 1,500-acre Neversink Reservoir.

We should begin by noting that the reservoir itself is an outstanding location for smallmouth bass, brown trout, rainbow trout, land-locked salmon, perch, and sunfish. When the dam was built to form the reservoir, more than 7 miles of river once famous for trout fishing were inundated. However, the reservoir compensates for this loss with cold releases of water that keep the trout fishing excellent for many more miles downstream.

The Neversink River is now, as it always has been, most noted for brown trout, although brook trout are found both above and below the reservoir. Perhaps the best trout fishing is found from the reservoir dam to the Route 17 Quickway bridge above Bridgeville. This long stretch of the river, cooled by releases of water from the reservoir and located in its entirety in Sullivan County, features a number of easements for anglers, which assure fishing access. To add to this accessibility, special anglers' parking areas are provided at no fewer than seven locations along this segment of the river.

Anglers are especially attracted by the large populations of wild trout that congregate between the reservoir and the area adjacent to the town of Hasbrouck. The wild populations are sufficiently large that the DEC does not stock these stretches.

However, trout are stocked farther downstream. In 1998, 6,500 browns—all of them 9 inches long—were stocked at Fallsburg and Thompson. (The river also received another 4,410 trout, including 550 that were 13.5 inches in length, downstream in Orange County.)

Anglers fishing between the two bridges in the Hasbrouck area are flanked by beautiful open country and some farmland. The quiet water and deep pools at the second bridge attract local anglers, who ply the waters with artificials and live bait. Among the wild trout are some exceptionally large fish that fire the imaginations of Neversink River anglers. Some feel, as does Jim Deren, angling philosopher and proprietor of a famous fishing lodge, that, "There don't have to be a thousand fish in a river; let me locate a good one and I'll get a thousand dreams out of him before I catch him—and, if I catch him, I'll turn him loose."

Tip: Stream browns are not the "epicures" some make them out to be. Successful anglers take them on salmon eggs, garden hackle, night crawlers, trout egg clusters, leeches, crawfish, minnows, crickets, hellgrammites, and kernel corn, to name but a few menu items. Of all these, perhaps the most effective are night crawlers, especially when fished close to bottom along banks, under logs, and near overhanging branches.

Although the effects of the cold-water releases diminish downriver, there are still plenty of trout in the lightly fished sections between

Fallsburg and the Route 17 bridge. Farther downstream, the river cascades into the spectacular Neversink Gorge, which cools the water in the river and nurtures trout populations. Significantly, the lands bordering the gorge were bought in the early 1980s by the state to create the Neversink Gorge State Unique Area, which includes a beautiful 3.5-mile stretch of river from Mercer Brook to the Sullivan/Orange County line. It is called "Unique" because of its unique rules and regulations, including prohibitions of fires, camping, rock climbing, swimming, horseback riding, launching of mechanically propelled vessels, snowmobiles, and removal or defacing of any state property, including rocks, fossils, plants, or trees, or molesting or disturbing any wildlife except during an open season.

Success rates for trout anglers is first rate in this cool, protected stretch of the river, whose interest is heightened by the occasional rattlesnake slithering through the rocks.

The fishery becomes considerably more diverse below the dam above Route 209. Shad swim up the Neversink and make it as far up as this dam, creating a fine shad fishery in spring in the Cuddlebackville vicinity. Bass and panfish are found in good numbers in the final stretch, from the Basher Kill down to the point where the Neversink enters the main stem Delaware River at Port Jervis.

35 Croton River, West Branch

Directions: Take Route 301 in Putnam County to the Boyd Corners Reservoir outflow in the town of Kent. Take Route 6 to the West Branch Reservoir outlet, and Routes 100 and 202 to the stretch below Croton Falls Reservoir.

The West Branch of the Croton River combines all of the qualities beloved by trout anglers with a passion for rivers and streams. The river is small enough to be comfortable. It runs cold and clean all year. It is a nursery for trout, and wild browns provide excellent season-long fishing opportunities. It is beautiful in all the ways that streams can be beautiful, tumbling through long, tree-shaded stretches of riffles, deep pools, and pocket water.

The West Branch emerges from Boyd Corners Reservoir and meanders a handful of miles to its confluence with the East Branch. Along its route, it connects three New York City (NYC) reservoirs. It begins as the outlet from Boyd Corners and flows a half mile to West Branch Reservoir. It begins again as the West Branch Reservoir outlet and flows 2.3 miles to Croton Falls Reservoir. A third section runs from Croton Falls to the East Branch. Consequently, a total of just less than 4 miles of river connects these three lakes and serves as a literal fish highway between them.

Boyd Corners is small by NYC reservoir standards at slightly less than 300 acres. The trout season on the short section of river emanating from Boyd Corners has classically been two weeks longer than the lower two stretches. The fishing is best in the spring, but anglers fishing in the last 30 days of the season to mid-October often catch larger fish, including large browns moving into the stream from the West Branch Reservoir. This stretch is stocked by the DEC (750 8.5-inch browns in a recent stocking year).

The longest of the three stretches begins as the outflow from the significantly larger West Branch Reservoir, which is just under 1,000 acres and is fed by streams that produce native trout. This section of the West Branch flows between tree-shaded banks and secluded woodland settings. This portion of the river serves as a spawning and nursery stream for Croton Falls Reservoir brown trout and consequently does not have to be stocked. Wild browns provide excellent season-long fishing.

The third and final section of the West Branch flows from Croton Falls Reservoir to its junction with the East Branch. Although it is only 1 mile long, this section straddles the Putnam–Westchester County lines. It is the most accessible section of the river, with entry points available from county Route 34 (Mahopac–Croton Falls Road) as well as N.Y. Routes 100 and 202. This section is regularly stocked by the DEC with yearling brown trout in the 8.5- to 9-inch size (1,350 browns were stocked here in a recent stocking season).

Special regulations along this final stretch of the West Branch make it especially attractive to anglers who fly-fish. This is a no-kill stretch that can be fished with artificial lures only.

Tip: This stretch enjoys good hatches plus banquets of terrestrials falling out of overhanging trees, providing excellent food resources for its wild and stocked browns alike. Be sure to figure it into your warm-weather trout-fishing plans. The cold-water releases from the Croton Falls Reservoir keep the fish feeding (and the fishing excellent) all season long.

It is remarkable that in less than 5 miles this wonderful little river can provide so much diversity and such extraordinary trout-fishing opportunities. It is not to be missed.

36 Connetquot River

Directions: Take the Southern State Parkway almost to its terminus at Hecksher Spur; then follow Route 27 east for a mile to the entrance to Connetquot River State Park.

One of the criteria we emphasized in selecting the 100 best fishing waters was that fishing is open to the public, because we want to present fishing opportunities that will be enjoyed by as many anglers as possible. However, our book would not be complete without reference to the Connetquot River, situated in the Connetquot River State Park in Suffolk County. Much about the Connetquot is unique in the fullest sense of the word, at least for American fishing venues.

The river begins in the north of the park, where springs gush clear, cool water year-round. The water flows over a bed of sand and glacial pebbles toward the river's mouth at Great South Bay. Along its route, it transverses some of the loveliest woodland on the South Shore of Long Island. And like that of other "spring creeks," its water temperature remains low and stable even in the hot summer months.

The fishing is fly-fishing only with barbless hooks, and access is by fee only. However, the trout fishing is so superb that anglers from all over the country travel to fish this little river in the center of Long Island. The park stocks 40,000 to 50,000 brook, brown, and rainbow trout, with fish coming from a hatchery located within the park. The largest number of stocked fish are hard-fighting brook trout. (There also is some natural reproduction of brookies in the upper stretches of the river.)

Each angler must purchase a "beat," a section of the river generally at least 100 yards long, which is his or hers alone to fish, just as anglers have bought and fished beats for generations in Scotland. There are twenty-three beats on the river itself, and nine more on the three ponds—Lower Pond, Main Pond, and Deepwater Pond—along the river. (The ponds should not be overlooked. The largest trout taken from the park may well be the 14-pound 10-ounce brown yielded by Main Pond.)

Each beat features a platform for casting; thus waders are not absolutely necessary. However, many anglers prefer wading, generally fishing downstream to the bottom of their beats, and then walking back to the top.

All varieties of trout will, when possible, head out to sea and return eventually in response to the spawning urge. It is therefore not surprising that many of Connetquot River's trout are sea-run, since Great South Bay beckons only several miles downstream. Tom Schlichter, the New York freshwater field editor for *The Fisherman*, told us that the best time of the year for sea-run trout is from February through April. In early spring, returning sea-runs will range anywhere from 2 to 20-plus pounds.

One of the key trout foods in the Connetquot is a little freshwater shrimp, which often is called "scud." In addition, the river has superb fly hatches, including brown stoneflies, quill Gordons, caddis, sulphurs, Hendricksons, and little green damsel flies, throughout the late spring and summer. Imitations of these natural foods are especially effective early in the season.

Tom's favorite all-around fly on the Connetquot is the size 12 dark Montana nymph. He added that many varieties of flies work well, with perhaps the best year-round fly a dark bead-headed nymph. He also favors a brightly colored salmon pattern.

Tip: Use bigger flies here, since many of the fish are quite large and prefer bigger meals when they head in to spawn. Try a Maribui Muddler Minnow in size 4 or 6, or a Giant Stone Fly in the same sizes. But don't forget that big breeders will occasionally rise to the smallest dry flies.

In warmer weather, fish begin to rise to hatches of caddis and diptera. Small Adams or pheasant-tail nymphs are good all-around producers in the summer. In the fall, zug-bugs and gold-ribbed hare's-ear nymphs are local favorites on the Connetquot.

Tom mentioned two other nearby locations (outside the park) where anglers can fish without fee and with bait, if they so choose, often with first-rate success. The spots, Bubbler's Falls and Rattlesnake Creek, are both located on the south side of Sunrise Highway and both also drain directly into the bay.

The Connetquot is closed to fishing through the winter months (November 1 to January 31), and of course access is restricted to those purchasing beats. (To reserve a beat, call 516–581–1005.) Special platforms are available on beats 16 to 20 and are reserved for persons who have physically limited abilities, including persons in wheelchairs. These are just several more aspects of the river that make it unique and thoroughly attractive.

37 Peconic River

Directions: Take the Long Island Expressway (LIE) to Route 24 at Riverhead. Follow Route 24 along upper portions of the river.

The Peconic River is another exceedingly pleasant surprise for anglers interested in quality freshwater fishing on Long Island. In its many miles of streambed and ponds above Riverhead, the river is chock full of freshwater species, most prominently largemouth bass, but also plenty of big chain pickerel, yellow and white perch, carp, catfish, and panfish. In its lower tidal flow, the river features huge numbers of (short) stripers, blues, and weakfish. Thus this magnificent river is a virtual buffet of fishing opportunity, freshwater and saltwater alike.

The Peconic starts out as a series of ponds and connecting brooks near Brookhaven National Laboratories and flows all the way to Gardiners Bay at the tip of the island. The quality of the water that feeds the river is excellent, and careful land-use controls and plenty of public education on preservation keep the Peconic among the cleanest surface waters in Long Island. Although this is a tidal river, it is regularly stocked by the DEC with brown trout in its upper reaches.

The upper river is dammed in several places, creating a series of ponds connected by sections of free-flowing river. The ponds offer good fishing opportunities, most especially Forge Pond. All of the factors are present in the upper river for nurturing warm-water species, including plentiful forage fish, cover, structure, and cool, well-oxygenated water. Consequently bass and pickerel growth is excellent and fishing is correspondingly excellent, with anglers often catching high proportions of keeper-sized fish. In the mid-1980s, a special "angler cooperator" program revealed that in one year, a surprising 60 percent of bass caught in one section exceeded the legal limit of 12 inches. Participants reported a lunker 22-inch bucketmouth and a 24-inch pickerel that year. And if anything, the fishing is even better today.

Many anglers use canoes to fish the river, especially in early summer and again in early fall, when plant growth is more restrained. Much of the best fishing takes place in a four-hour canoe stretch from Edwards Avenue (near the LIE overpass) to and through Forge Pond. Crankbaits, stickbaits, spinnerbaits, and rubber worms are preferred artificials, but plenty of largemouths are taken on a variety of live baits as well. In this especially beautiful stretch of the river, the method becomes almost academic when viewed against the sheer pleasure of the enterprise. Author John Buchanan described that pleasure well when he wrote, "The charm of fishing is that it is the pursuit of something that is elusive but attainable, a perpetual series of occasions for hope."

The excellence of Peconic River fishing is complemented by its convenience. Boats are protected in the river, so when winds blow up rough seas in the bay and ocean, anglers can still go fishing here for saltwater species. Moreover, there are plenty of shoreline locations where anglers can stand and cast successfully for stripers and blues. A good place to start, especially if you're taking the kids, is the dock in downtown Riverhead, where snapper blues start to bite in August and short stripers and weakfish are also taken in numbers.

Downriver, a real hot spot for anglers in boats is the area in and around the Route 105 bridge.

Tip: Remember that stripers often circle in tight schools and can be found almost anywhere. If you jump a fish when trolling or even drifting, mark the spot carefully and come back to it as quickly as possible, casting the same lure that attracted the first fish.

Fly-fishing for saltwater species on the Peconic estuary is a real possibility. Some of the charter boats carry flyrods on board as standard equipment. Recommendations for taking stripers include big 3/0 streamers with fluorescent red hackle collars and white saddle hackle tails. For the rest of us, one of the truly exciting ways to take bass is livelining alewife or blueback herring, noted to be high on the preferred menus of linesiders large and small.

Regarding boating access, the DEC maintains small hand-launch sites on the upper Peconic in Calverton on South River Road, and on the lower Peconic off N.Y. Route 25 in Riverhead. Upper river or lower, largemouth bass or striped bass, you cannot make a bad decision on the Peconic.

Reservoirs

Water for New York City is impounded in three upstate reservoir systems that include eighteen reservoirs and three controlled lakes with a total capacity of 550 billion gallons. The reservoirs range upward in size from Lake Gilead, the smallest at 121 acres, to Ashokan Reservoir, the largest at 8,315 acres; and in depth from Diverting Reservoir, the shallowest at 34 feet, to Rondout, the deepest at 190 feet. All share certain characteristics that are of specific interest in the context of this book. All have wonderfully clean and productive waters. All are administered under an extraordinary set of regulations calculated to keep them as pristine as they currently are. And all are loaded with trout, bass, and panfish.

38 Cannonsville Reservoir

Directions: Follow Route 10 in Delaware County along long stretches of the northern and southern shorelines.

Cannonsville Reservoir combines features that should warm the cockles of any angler's heart. It is heart-stoppingly beautiful. It is very lightly fished. And it is home to large and voracious game fish in numbers, including brown trout exceeding 20 pounds.

Created when the West Branch of the Delaware River was dammed in 1967, the slender reservoir is some 15 miles long, averaging only 2,000 yards in width. The longer of the reservoir's two upper branches reaches up the West Branch river valley toward Beers Brook. The shorter floods up the Trout Creek course. The reservoir supplies water to New York City via the West Delaware Tunnel. Interestingly, cold water also is released into the West Branch of the Delaware, creating a fine trout fishery in the river below the dam.

By late fall, drawdowns of water drop the reservoir's level by as much as 60 feet. However, fall rains refresh and refill the reservoir, so that it is generally full again by early spring.

Many species of fish—especially trout and smallmouth bass—grow at startling rates on shoals of alewife herring in the reservoir. The upper stretches of the reservoir are significantly affected by the heavy nutrient flow from the West Branch, which nurtures oxygen-depleting algae in warm weather. Consequently trout anglers work the upper 4 miles of the reservoir in the warm months, where deeper, colder water, which is relatively free of algae, provides conditions that sustain trout. Significantly, trout scatter the length and breadth of the reservoir in cooler weather.

Smallmouth bass weighing 4 to 5 pounds are not unusual, and given abundant forage fish and light fishing pressure, their numbers continue to grow. Yellow perch in deeper water and big carp in the shallows also provide great fishing fun.

Tip: The most convenient bait for Cannonsville carp is thoughtfully prepared for anglers by the Green Giant. Kernel corn threaded onto a #4 or #6 baitholder hook and fished at bottom will produce as many carp as any other bait in our experience. Prepared carp bait and homemade cornmeal bait, flavored with everything from anise to onion to vanilla extract, also works quite well.

In his *San Juan River Chronicle*, angler Steven J. Meyers wrote that, "No trout, except possibly a very old, very heavy, very wise trout, fights like a large carp." For Cannonsville Reservoir anglers out for a morning of pure fun, this observation should not be taken lightly. Indeed, we have never found the trout that equals a large carp in strength or tenacity.

39 Pepacton Reservoir

Directions: Take N.Y. 206 west from Roscoe to the intersection with Route 30 at Downsville and the reservoir. Route 30 closely follows half of the reservoir's north shoreline from Downsville to the Shavertown Bridge, then crosses the reservoir and follows its south shoreline the remainder of its length.

Discussions of most fine fishing waters in New York State require at least a nod in the direction of a number of fish species, especially various strains of bass, salmon, and trout. However, the exceptional fishing in Pepacton Reservoir appropriately concentrates on one fish alone: the brown trout. Not that other fish do not exist in this massive body of water. Given that the fish-rich East Branch of the Delaware River is its principal feeder stream, the reservoir clearly has many species present. However, among all of the lakes and reservoirs in the region, Pepacton is considered by many the best for brown trout.

The DEC's excellent stocking program in the area speaks volumes about the brown trout in Pepacton. In a recent stocking season, the East Branch was stocked at sixteen different locations in Region 4, and every fish stocked was a brown trout. The smallest of these fish were 8.5 inches in length, the largest 14 inches. These trout do well in the cold, clear water of the East Branch of the Delaware; however, when they swim into the reservoir, with its superb habitat and plentiful stocks of sawbellies (alewife herring), they begin growth spurts that can be extraordinary.

Significantly, however, wild fish comprise the major component of the reservoir's trophy trout fishery. DEC tagging studies show that small, wild browns from a variety of feeder streams migrate down to the reservoir to become the giants patrolling its depths. Growing rapidly on herring, the browns reach 8 pounds in five years, and to 20 pounds or better if they survive another four years. The average trout taken in the reservoir is 12 to 18 inches long, weighing from 1 to 3 pounds. However, browns up to 21 pounds have been taken, and beyond question there are still larger trout in the lake.

Pepacton provides more than sufficient room for these great fish to grow. Created in the mid-1950s by impounding the East Branch of the Delaware, the reservoir is roughly 20 miles long and .5 miles wide, with some 5,700 surface acres when water capacity is at optimum level. (Gradual drawdowns in summer and fall generally lower the lake level by 20 to 30 feet.) The water quality is so good and the dissolved oxygen so well distributed that browns have been taken at bottom in the deepest part of the lake, 160 feet down.

Slow drift fishing with live bait worked in the thermocline probably catches more large browns here than any other method. As with the other New York City reservoirs, special regulations apply. Boating permits must be secured (and current) and boats must be left locked at lakeside. Outboard motors are prohibited. (Thus it is good to remember that a long drift in a favoring wind will produce fine trout sometimes, but will yield a long row back to your starting point every time.) Another significantly productive (though less-often discussed) method for taking big trout is double-anchoring over a good drop-off where fish are marked electronically, working live bait on flat lines, at preferred depths on slider floats or suspended. Depending upon the season, water temperature, light conditions, and other environmental factors, thoughtful presentations of bait from a carefully double-anchored boat will produce fish.

Tip: The correlation between water temperature and successful fishing in Pepacton and other reservoirs is almost absolute. Temperatures between 55 and 65 degrees will yield the most brown trout, a fact that helps determine locations to fish with the passing of the seasons. For example, in mid-April, temperatures in the reservoir are generally constant from bottom to top: about 42 degrees. However, by mid-June, the top 15 feet ranges upward from 60 to 70 degrees, an uncomfortable range for browns. The layer from 15 feet down to about 30 feet will be perfect, going from 55 to 68 degrees, with the remaining water to the bottom ranging down from 55 to 42 degrees. By August 15, the top 25-foot layer will be too warm for browns, with the water from 25 feet down to 35 feet well in the trout "comfort zone" of 55 to 68 degrees. Naturally, the thing to do is (1) be continuously aware of water temperatures, and (2) fish the layers where the browns are most active. Fishfinders are often critical to this effort.

Two factors combine to equal good shore fishing for browns at Pepacton in the early spring. First, water temperatures at or near the surface are ideal for feeding trout. Second, there are many steep drop-offs within easy casting distance of shore, where good fish work the edges.

We noted earlier that other species of fish are available in Pepacton. In fact, there is a fine smallmouth bass fishery in the reservoir, which largely escapes notice of trout-fascinated anglers. The smallies are big, averaging between 1.5 and 3 pounds, and occasionally running up to 6 pounds or more. The most effective method for catching these fine bronzebacks is casting live bait—preferably crawfish—from boat to rocky points, drop-offs, and escarpments along the shoreline. Carefully worked sawbellies, lip-hooked and lightly cast to the rocks, also produce bass, as do a wide variety of artificials, ranging from crankbaits that imitate crawfish to grubs. Prime smallmouth habitat is to be found near the dam and at the mouths of certain tributaries, and especially in the spring, the action often can be hot.

Boat access is limited to those who hold permits or who park rowboats lakeside on a permanent basis. Access is good for shoreline fishing, with information, live bait, and even a guide or two available at local tackle stores.

40 Ashokan Reservoir

Directions: Take the New York State Thruway (87) to exit 19, to Route 28 west. Route 28 circles the entire northern expanse of the reservoir to Boiceville, intersecting with Route 28A at Boiceville. Route 28A parallels the reservoir's southern shoreline.

The sensation at first glimpse of Ashokan Reservoir is that one has found an inland sea. The largest of the Catskill reservoirs by far, the Ashokan covers 8,300 acres and is, at its widest point, well over 2 miles across.

Figure 13 ■ Ron Bern and Manny Luftglass fish Ashokan Reservoir as part of a field trip.
(Photos: The authors.)

The Ashokan is regularly mentioned by anglers who ply the trout-rich waters of Esopus Creek, which is the trout hatchery for the Ashokan. The creek's extraordinary populations of trout—especially rainbows—migrate downstream to the reservoir at about one to two years of age. Rainbow and brown trout grow steadily on the Ashokan's significant populations of alewife herring ("sawbellies" in local parlance) and emerald shiners, then return to the creek at about three years of age to spawn. By this time, the spawners have reached trophy size, between 16 and 22 inches, and are much prized for their power as well as their girth.

The reservoir is essentially two reservoirs, divided by a dam, or "weir" on Reservoir Road between Shokan and Olivebridge. The upper (west) basin is somewhat the smaller of the two, covering roughly 3,000 acres, and also the deeper, with depths exceeding 140 feet. The lower basin, whose 5,000-acre expanse reaches back toward the Thruway, reaches depths of up to 90 feet. The two basins are distinctly different, with the lower being somewhat clearer because the upper serves as a "filter" for the relative turbidity of Esopus Creek. However, both basins hold fish in common. Fish regularly make their way over the weir from upper to lower basin. (The proof is the presence of rainbows in the lower part of the lake, since "bows" are not stocked at all in the Ashokan.) When conditions are right, trout can be seen jumping the weir in the opposite direction as well.

If brown and rainbow trout are the most celebrated denizens of the Ashokan, they only begin to tell the story of this massive, beautiful body of water. Fine smallmouth bass, walleye, carp, and yellow perch also regularly come to creel in the reservoir.

One of the hottest spots for big walleyes is the legendary Chimney Hole, which is in fact a large pool at the juncture of the upper reservoir and Esopus Creek. This hole, which acts like a magnet for serious anglers intent on big fish, is near the top of our future "must fish" list for the authors.

The lower basin, more than the upper, is affected by drawdowns of water for New York City. Indeed, when we fished the lower basin together, the water appeared to be down at least 30 feet. Noting the high water line at the base of a stand of trees, we walked on perfectly dry land for more than 150 feet before unlimbering our spinning rods at water's edge.

Tip: Trout anglers will want to be aware of water temperature, especially in warm weather. The thermocline goes progressively deeper as New York City draws off the colder bottom layer of water. Anglers equipped with thermometers will find both the thermocline and most of the trout especially deep in July and August.

"Row trolling" is a popular if strenuous method of fishing the Ashokan, with slow-trolled spoons especially popular with trout anglers. Shore anglers take big walleye at night on Rebels and Rapalas and use the same artificials for brown trout and rainbow trout during the day. November and April are the top months for fishing from shore, especially at the weir. Sawbellies and emerald shiners (locally called "icesickles") are the deadliest natural baits.

Fishing access is excellent for anglers with the requisite fishing permits, which are issued at no cost by the New York City Department of Environmental Protection (DEP). As in the other New York City reservoirs, no motor boats are permitted on the water and car-topped or trailered boats are prohibited. As on other reservoirs, boats used on the Ashokan must remain chained to a tree on shore, must be permitted, and must be inspected at the mooring site each year to be certain the boat is clear of zebra mussels.

Requirements and strictures notwithstanding, this vast, beautiful lake is one that should not be missed, especially when trees are greening in spring and when they are turning in the fall.

41 Rondout Reservoir

Directions: Rondout lies astride the Ulster and Sullivan County lines. Take Route 55 or 55A along the shorelines of the reservoir.

Rondout Reservoir is one of the deepest of the Catskill reservoirs, which figures significantly in the reservoir's wonderful fishing opportunities. The reason: a large, naturally reproducing population of lake trout thrive in its deep, cold water.

Rondout is a long, narrow lake, some 6 miles in length and a half mile wide at its widest point. Roughly rectangular in shape, it reaches depths of 189 feet, and its mean depth is almost 74 feet. It is fed by two underground aqueducts and by noted trout streams, including Chestnut Creek, Trout Brook, and the upper Rondout Creek.

While anglers ply the depths for lakers, they also fish the reservoir for brown trout, which have been stocked here for many years and which also make their way into the reservoir from feeder streams. Some anglers prefer to rowboat-troll flutter spoons in blue and silver colors that imitate alewife herring, the predominant forage fish. Others have success casting from shore with lures or spoons of the same color. The shoreline of Rondout has steep drop-offs and plenty of clear shoreline, making casting from shore a feasible alternative to boat fishing. (At this writing, brown trout must be at least 12 inches in length, and lakers 18 inches, to be kept; three of each may be legally creeled. Of course, rules change, and a careful reading of the compendium is highly recommended.)

Tip: Although lakers will readily take shiners, sawbellies are their natural food fish in the lake and are thus the primary baitfish. However, sawbellies are not available in colder weather. Thus it is a good idea to freeze unused herring from earlier fishing trips and bring them along on ice. Some frozen herring will float in your bait pail while others will sink to the bottom. Try the RB method of hooking a floating herring about halfway between tail and dorsal fin and dropping it deep with a three-quarter-ounce egg sinker resting on bottom. The herring will "swim" above the sinker in a natural-looking upright attitude, thus attracting lake trout that would be put off by an unnatural presentation.

Rocky structure along the shoreline provides habitat for smallmouth bass, which are taken on both artificials and carefully presented live bait. Sawbellies are best for bronzebacks, although crayfish, leeches, and night crawlers produce catches as well. Other fish present in numbers include white suckers running up to a pound and significantly larger carp.

Ice fishing is not permitted on Rondout or other New York City reservoirs located west of the Hudson. Trout and bass seasons are closed during the winter and the reservoir is not noted for pickerel or perch at any rate.

The quality of fishing in Rondout is a pleasant surprise, given drastic reductions in all species after heavy copper sulfate and chlorination treatments in the 1980s. In fact, the reservoir has rebounded exceedingly well, due in large part to excellent stocking and fishery management efforts by the DEC (3,400 lake trout and 5,300 brown trout in the 8- to 9-inch range were stocked in one recent stocking year).

In 1991, one thoroughly surprised angler took a 12-pound landlocked salmon in the reservoir. On consideration, it is likely that the salmon entered the reservoir via the tunnel from Neversink River to Chestnut Creek.

Access is controlled by regulations associated with all of the New York City reservoirs, including required fishing permits and rowboats (without motors) that must be permanently stored in approved areas. However, the lake trout fishing alone at Rondout makes these minor inconveniences more than worthwhile. The brown trout and smallmouth bass fishing are significant bonuses.

42 Swinging Bridge Reservoir

Directions: Follow Route 17B west from Monticello to Starlight Road exit. Starlight Road traces most of the eastern shoreline of the reservoir.

Swinging Bridge Reservoir lies in the Mongaup Valley in Sullivan County, between the smaller Sackett Lake to the east and Lebanon Lake to the west. Fed by the Mongaup River, the reservoir is upwards of 9 miles long, with isolated holes approaching 100 feet in depth and widely varying habitat conditions, ranging from numerous rocky points, drop-offs, and old foundations to shallow, weedy water in the upper coves and bays. Not surprisingly, this all spells bass with a capital "B." Indeed, Swinging Bridge has been called the finest smallmouth bass water in the region, and not without cause. Especially in the reservoir's steep lower end, where gravelly bottoms and rocky points abound, smallmouths in the 2- to 3-pound range provide wonderful sport for anglers working live bait and artificials.

Significantly, anglers set for the smash of one of these scrappy bronzebacks are sometimes shocked by a powerful lunging run that either breaks their monofilament like a thread (if the drag is set a bit too tightly) or tears off a hundred feet of line on an open drag in the blink of an eye. And if the drag is set properly and the fish is hooked, the angler should figuratively fasten his or her seatbelt for a spectacular fight, courtesy of the DEC, who selected Swinging Bridge Reservoir as an experimental site for stocking hybrid bass.

We must confess here a distinct prejudice in favor of this spectacular hybrid, which combines the genes and best qualities of the striped bass and the white bass. Pound for pound, we believe the hybrid bass fights like no other freshwater fish that swims. Some species may strike lightly at one time and more powerfully at another time; other species may fight savagely just after the strike and less powerfully as the fight wears on; still others come to the boat fairly easily, saving energy for a driving lunge right at the net—but the hybrid bass will have none of this. Like a waterfront bare-knuckled brawler, the hybrid shows up to fight, from the moment the fish attacks the bait until he either breaks your line or fights himself into a state of exhaustion. The speed of his run is dizzying and the quality of his fight is sufficiently exciting to make converts of lifetime trout and smallmouth anglers.

Tip: Hybrid bass show a marked preference for sawbellies, generally of midrange size, at various depths from surface down to 30 or 40 feet, depending upon wind, water temperature, and light. A flat-line (that is, a line without sinker or float) with a good-sized sawbelly working at his own comfort level is a good bet for hybrids. A second line rigged with a slider float and worked well away from the boat, or a suspended line with a sliding three-quarter-ounce sinker worked at a depth where you electronically mark fish, also are good bets. More than incidentally, hybrids prefer lively, unmarked bait. And while they certainly are taken on lures that imitate sawbellies in appearance and action, we believe that in the long run they're vastly more susceptible to the "real thing."

Note: There is not an inexhaustible supply of hybrids in Swinging Bridge, since (1) this fish seems ingenious in its ability to escape lakes and reservoirs, (2) hybrids have not been restocked, and (3) being hybrids, they cannot reproduce. However, there is a remnant fishery in the reservoir and a number of fish as large as 10 pounds still await your offering.

Beyond question, Swinging Bridge is best known for bass. However, there also is a thriving fishery of brown trout and walleyes. The trout are essentially refugees from the Mongaup River, the principal feeder stream to the reservoir, which contains wild browns whose populations are regularly augmented with DEC stockings above the reservoir. Every year, bragging-sized browns are taken in the reservoir, some as large as 7 pounds. The new walleye fishery is carefully managed, with tight creel limits in place (three fish 18 inches or longer at this writing). The DEC stocks large brown trout and numbers of fingerling walleyes in the reservoir. In a recent stocking season, 20,000 small walleye were added to this fine reservoir and upwards of 100 trout ranging from 17.5 to 22 inches.

At this time, there is no limit on the sizes of boats or motors. Public access is guaranteed, but boat launching is limited to an unimproved dirt ramp maintained by the power company that controls the reservoir, or a paved pay ramp at Swinging Bridge Marina, located at the midpoint of the lake's eastern shoreline—both reached via Starlight Road. This limited access represents a small inconvenience. However, we assure you—indeed, we guarantee—that a single thrilling hookup with a hybrid bass, even if you never land the fish, will be more than sufficient compensation.

43 Titicus and Muscoot Reservoirs

Directions: Follow U.S. Route 684 up the Croton River in Westchester to Purdys. Exit at Route 116 east at Purdys and follow a short distance to the Titicus.

The Titicus and the Muscoot are among the most appealing of the New York City reservoirs, not only because of their natural beauty but also because of their oversized game fish. The two reservoirs are considered together here because they are connected by a short section of the Titicus River (reservoir outlet), which flows for only a half mile from the Titicus Reservoir into the Muscoot. Both rainbow and brown trout are stocked in this stretch of river, which is fast-flowing, very accessible (from N.Y. Route 22) and a first-rate trout fishing venue itself. However, the two reservoirs it connects are considerably more interesting for our purposes.

Titicus Reservoir has a shoreline of slightly more than 8 miles and comprises just over 680 acres of surface area. It is nearly 80 feet at its deepest point—a hole located near the dam—and 60 to 70 feet deep in the area from its most prominent island to that hole. This deeper water is critical to the thriving brown trout population, which is regularly replenished with stockings of yearling fish by the DEC. (In a recent stocking season, 3,000 9-inch browns were put in the reservoir at North Salem.) Titicus is fed by the Titicus River, a small, rocky stream with many deep pools. This inlet stream has its own trout fishing possibilities and is simply another attractive part of the Titicus-Muscoot system.

The most impressive denizens of the Titicus, however, are largemouth bass that grow to proportions more usually encountered in the deep South. In November of 1992, for example, local angler Tom Schaub caught an 8.75-pound "bucketmouth" in the reservoir, and it is likely that even larger bass still swim in the weedy shallows of its back coves. The reservoir also has a fair number of smallmouths in its widely dispersed rocky formations, plus substantial populations of perch, both white and yellow.

The Muscoot, which covers 1,263 acres of surface area, was famous in years past for incredibly large and powerful tiger muskellunge (a rapidly growing cross between northern pike and true muskellunge). The DEC stocked the fish in 1980 and steadily larger fish were taken, up to and including a 16.75-pound monster wrestled out of the water in 1984. However, that stocking program was discontinued because tiger catches were just too few and far between. The reservoir is basically shallow and weedy—in other words, perfect largemouth habitat. Anglers fish virtually the entire reservoir for bass, favoring the areas in the northern quadrant for largemouths and the deeper portions of the lake at its southern end for bronzebacks.

Since smallmouths love structure, it is worth noting that the Muscoot is littered with stone walls, old foundations, and of course, big stone abutments where bridges cross the reservoir. Working the submerged structure and abutments with spinners—especially smaller silver or bronze Mepps—and the usual varieties of live bait will put smallmouths on the scoreboard.

Tip: On impoundments, bridges normally run along creek channels, which make them signposts to the deepest water in the area. Bass use creek channels as natural highways and when they encounter bridges, they tend to stay because of cover, structure, shade, and in some cases, current. When fishing bridge abutments, position your boat so that you can cast and retrieve lures across points and parallel to pilings. Experiment with lures to determine productive depth, starting high with spinnerbaits and curly tailed grubs and working deeper with crankbaits and Texas-rigged worms.

The Muscoot is fed by many trout-holding tributaries; consequently, stocking by the state is not required to keep trout in the reservoir (except for the aforementioned stocking of rainbows and browns in the feeder Titicus River). However, the Muscoot is generally shallow, becoming quite warm in the summer. Thus anglers seeking trout would be better advised to visit nearby Kensico, Titicus, or Croton Falls Reservoirs. The Muscoot also supports wonderful populations of panfish, most prominently big schools of crappies. The standard crappie rig is a small float, split shot, and minnow-baited hook. However, if throwing artificials, keep them small and shiny and retrieve slowly. And keep your line tight after setting the hook, since hooks easily tear loose and fall out when lines go slack.

The Muscoot is reached by exiting U.S. Route 684 at Route 35, which crosses the reservoir at its constricted midpoint near Whitehall Corners. All of the usual regulations for New York City Reservoir System apply in both reservoirs.

44 Croton Reservoir

Directions: Take Route 9A up the Hudson to Route 129 at Croton-on-Hudson. Follow 129 a short distance to and around a portion of Croton Reservoir.

Croton Reservoir has to be experienced to be believed. Lying in populous Westchester County, only a stone's throw from the busy highways that track the Hudson River and the bustling community of Croton-on-Hudson, the reservoir is nonetheless unspoiled, remote in every aspect except geography, and filled with scrappy game fish.

We should note that the name of the reservoir sometimes causes a bit of confusion. Technically, the name now is the New Croton Reservoir, as it is represented on newer maps. However, anglers still tend to refer to it by its older, simpler name. It is not to be confused with Croton Falls Reservoir, which lies a considerable distance to the north. However, they are both parts of the Croton Watershed System, which consists of twelve reservoirs, including the Croton (New Croton), Muscoot, Titicus, Cross River, Amawalk, Diverting, East Branch, Bog Brook, West Branch, Middle Branch, Croton Falls, and Boyds Corner, plus three controlled lakes (Kirk Lake, Lake Gleneida, and Lake Gilead).

Significantly, the water from the northern reservoirs flow via natural streams to the downstream reservoirs, terminating at the Croton Reservoir, which delivers water to thirsty New York City via aqueduct. Of more immediate interest here, of course, is the fact that Croton is continuously renewed and refreshed by waters and fish from the upstream reservoirs.

The reservoir originally was created by damming the Croton River and flooding upstream valleys. It is the most irregularly shaped of all the New York reservoirs, long and slender with one arm extending north by east almost to the Muscoot Reservoir. In all, the reservoir covers 2,304 acres.

Fishing pressure in the Croton is relatively light, perhaps because of its location so close to civilization. However, the bass fishing is nothing short of sensational, with skilled anglers regularly taking fine largemouths and smallmouths. As in most impoundments, the largemouths prefer shallower, weedier bays and creek mouths, while the smallmouths concentrate around rocky bottoms, ledges, and structure. The northernmost section of the impoundment where Hunter Brook feeds into the reservoir is rocky and filled with shoals and drop-offs, and is arguably the single best spot for bronzebacks. The southerly arm of the lake bordered by Route 134 is shallow, with plenty of bays and weedlines, and yields more than a fair share of largemouths. Anglers take both species of bass on a variety of artificials, especially artificial worms and crank baits. Big minnows and sawbellies produce well in the weedlines and along rocky drop-offs, and of course, the ever-dependable crayfish and hellgrammites are wonderful baits for smallmouths.

Tip: Another baitfish that produces well here is the killifish (killie), millions of which swim in the nearby Hudson River. With a bit of practice with a cast net, anglers can catch bucketsfull of these hardy forage fish, and score big on both largemouths and smallmouths at Croton Reservoir.

While most bass anglers believe that the smallmouth fights better ounce for ounce, there are those to whom the "bucketmouth" is the pinnacle of freshwater fishing excitement. As John Alden Knight writes in his book, *Black Bass*, "The honest, enthusiastic, unrestrained, wholehearted way that a largemouth bass wallops a surface lure has endeared him forever in my heart. Nothing that the smallmouth does can compare with the announced strike of his big-mouthed cousin." The 5- and 6-pound largemouths that work the shallows in Croton Reservoir may very well make converts to this point of view.

Bass anglers take more than a few pickerel on live bait, and healthy populations of panfish—especially crappie, yellow perch, and white perch—make Croton Reservoir a superb place to take the kids. Catfish grow to good proportions here, too, and big carp prowl the shallows, waiting to be tempted with cornmeal bait, corn kernels, or commercially prepared baits. (As with all other New York City reservoirs, special permit and boating regulations apply.)

The outlet that flows just over 3 miles from the reservoir to the Hudson River is the final section of the Croton River and is stocked with browns and rainbows by the DEC. Access to the first mile of the stream below the dam is available via Croton Dam Road and N.Y. Route 129 and fishing for both stocked fish and holdovers (following summers with good water conditions) is often outstanding. This is simply one more pleasure to be found at this "remote" lake that is surprisingly close to civilization.

45 Kensico Reservoir

Directions: Take Route 684 in Westchester County to Route 22 south, which crosses the foot of the lake.

Kensico is one of the most interesting reservoirs in the New York City water system. With a shoreline of nearly 22 miles, the irregularly shaped reservoir comprises 2,145 acres. The reservoir is constricted at a center point where it is crossed by Rye Bridge. The larger, deeper section stretches from Mount Pleasant to Valhalla. The smaller section—sometimes referred to separately as Rye Lake—wraps around a large island. Its habitats range from long stretches of shallow, rocky shoreline to holes as deep as 130 feet. With its exceptionally clean water, Kensico is a nearly perfect environment for nurturing trout, bass, and panfish.

The most southerly of the local watershed reservoirs, Kensico was once prized primarily for its lake trout. With its resident populations of sawbellies and its significantly deep water, the lake nurtured natural and stocked lakers to surprising sizes. (More than a few lakers in excess of 20 pounds were taken in the lake.)

However, the DEC switched its stocking emphasis from lake trout to brown trout several years ago because lakers reproduce naturally in the deep water of the reservoir. Last year, for example, DEC stocked 10,000 9-inch brown trout in the reservoir at Harrison, followed by another stocking of 3,000 7-inch browns. The majority of trout taken now are browns, primarily because few anglers know or use the specialized techniques needed to take lake trout consistently.

Unlike lakers, the browns do not naturally reproduce in Kensico, since they require a significant tributary stream for spawning. However, they grow quite quickly on a rich sawbelly diet.

Figure 14 ■

Danio Vaknic, Jim McGinnity, and Joe Valentinetti with brown trout at the Kensico Reservoir.
(Photos: Joe Valentinetti.)

Tip: *Lots of browns are taken from shore, especially early in the season. When shore fishing, try "pop-up" rigs combining a small marshmallow and a baby night crawler. The marshmallow will serve two purposes: first, to attract trout, and second, to float your nightcrawler up off bottom so that it is more visible to feeding trout.*

Kensico has a year-round trout season, although ice fishing is prohibited. Some of the best lake trout catches are made by anglers spin fishing from shore in November using Krocodyles or Little Cleos in silver or blue and silver (.5 to .75 ounces in size).

The bigger browns generally are taken by row trolling or drifting artificials (preferably flutter spoons) and live bait, or by fishing sawbellies via "sinker-free" flatline, slider float, or suspended lines from anchored boats. (Our preference here is a double-anchored boat for optimum control.)

Black bass also grow quickly in Kensico. Because of habitat, smallmouths outnumber largemouths, since there are many more rocky stretches, both at bottom and along the shoreline, than weedbeds and other "soft structure" preferred by "bucketmouths." Regarding bottom structure, a good fish finder is a key to locating the structure that will harbor smallmouths in the shallower sections (and lakers in deeper water).

As noted, large lake trout have often been taken in Kensico, and current populations of browns and black bass also yield fish of gratifying proportions. By the same token, several other species produce outsized fish here. For example, live bait intended for bass and trout often is taken by exceptionally large pickerel or yellow perch. In fact, some believe the largest pickerel in any of the watershed lakes swim in Kensico. And yellow perch as large as 16.5 inches have been taken here.

According to DEC's Bob Brandt, "Two NYC aquaducts enter Kensico: the 'Catskill' Aquaduct off Nanny Hagen Road at the northern tip of Kensico Reservoir and the Delaware Aquaduct at Dark Hollow in Rye Lake. During summer, these two cold-water inflows have a greater influence on the species and distribution of fish than water depth."

Bob adds that there also is a population of ciscos, which provide a popular early spring fishery just after ice-out and advises trying tiny minnows or long, narrow lures such as a Swedish Pimple. A popular spot for ciscos is near the mouth of the Catskill Aqueduct off Nanny Hagen Road.

As in all of the New York City reservoirs, only boats with special boat numbers and permanently remaining on reservoir property may be used, and anglers require valid NYS fishing licenses and reservoir permits.

And of course, no motors of any kind can be used. However, in many of the wonderful lakes in this system and most especially in Kensico Reservoir, these are but minor inconveniences when compared to the wonderful fishing for big, exciting game fish so close to major population centers.

Lakes and Ponds

Among New York's 4,000 lakes and ponds are some of the most beautiful, productive, and unusual still waters in the country. These range from a little sweetwater lake surrounded on all sides by saltwater in Montauk to the vast 51,000-acre Lake Oneida, the largest inland lake in New York; from the unique St. Regis canoe ponds to the vast Lake George. We present a small sampling here.

46 Black Lake

Directions: Exit N.Y. Route 812 at Gouveneur; take Route 58 to a crossing at the midpoint of the lake and the Hammondsville-Edwardsville County Road for shoreline access.

Black Lake in St. Lawrence County is often referred to as an angler's dream, and with good cause. Covering an area of almost 8,400 acres, this 20-mile lake is fed by the Indian River and runs to a maximum depth of 29 feet. It lies cheek by jowl to the St. Lawrence River and features irregular shorelines, mid-lake shoals, rocky points, expansive weedbeds, and many islands throughout its long, slender configuration. These in turn provide favorable habitat for thriving populations of a wide range of game fish, ranging from bass and northern pike to exceptional populations of black crappie, a newly regenerated fishery for walleyes, and even the occasional sturgeon.

The lake is perhaps best known for bass. Indeed, the expansive largemouth and smallmouth bass populations regularly attract bass fishing tournaments to Black Lake. In addition, log-length northerns are regularly taken in spring and through the ice in the dead of winter.

However, bass and big pike only begin to tell the lake's story. Large populations of big black crappie provide superb year-round fishing action for anglers who travel to the lake from all over the Northeast. So consistent is this calico bass fishing that some consider it the best in New York State.

Tip: Unlike white crappies, their first cousins, black crappies favor hard bottoms and abundant vegetation (over sunken brush, trees, and other underwater cover). The nickname for the crappie, "papermouth," is not accidental; it refers to a paper-thin membrane around its mouth. Do not reset the hook or pump the rod when a crappie is hooked. This may occasionally be indicated for other game fish; however, it will tend to tear the hook right out of the crappie's delicate mouth.

The lake also is loaded with other panfish, prominently to include big yellow perch. Moreover, although they are protected, a growing population of sturgeon feed in the forage fish-rich waters of Black Lake, growing out of DEC stockings that began in September of 1995.

The biggest news in Black Lake, however, is the return of the walleye. Biologists from DEC Fish and Wildlife decided that conditions warranted a significant stocking effort, and now tens of thousands of fingerling walleye, from .5 to 4.5 inches, are released in the lake each year, usually at Hammond and Morristown. The state also provides walleye fry from the Oneida Hatchery to volunteers of the Black Lake Fish and Game Club, who rear these to fingerling size and distribute them in the lake. Survival rates appear excellent and anglers are catching and releasing many undersized fish, the surest sign the stocking program is working.

The fishing is year round at Black Lake, with first-rate ice fishing for crappie, northerns, and perch.

Access to the lake is excellent and a fine concrete boat-launch ramp is maintained by the State Parks Department adjacent to the Hammonds-Edwardsville County Road, 2 miles west of the hamlet of Edwardsville in the town of Morristown. The site provides parking facilities for 55 cars and trailers.

47 St. Regis Canoe Area

Directions: Follow N.Y. Route 30 past the Upper Saranac Lake to Little Green Pond near the state fish hatchery.

Almost a third of a century ago, *Field and Stream* magazine featured an article about a trip into the "pack-in ponds" of the Adirondacks. The author had begun the trip into roadless reaches of the mountains with the conviction that serious trout fishing in New York State was a thing of the past. What he found, however, was a wilderness adventure in which gorgeous, virginal waters teemed with wild brook trout, and the only other anglers ever in sight or hearing were one's own companions.

"There are spots that haven't changed in the past hundred years or more, and some have been revived and are better than ever," the author reported. "But you'll never be able to drive to them in a car; that's what's saved them. Oh yes, there's great fishing still left here, but you have to work for it."

These conditions remain the same to this day, by virtue of superb efforts by New York State to maintain this unique wilderness experience. In the early 1970s, when all-terrain vehicles and snowmobiles threatened the tranquillity of even these remote regions, the Official State Master Plan for Adirondack Park, approved in 1972, established the new zoning category of "Canoe-Only Areas." The St. Regis Area was the first to be so designated, and the consequences have been all but magical.

There are some 300 ponds in the Adirondacks. About 20 of them form the headwaters of the famous trout stream, the West Branch of the St. Regis River. These range in size from just over an acre to more than 300 acres, and reaching them still involves carrying in canoes, food, and gear. There are two traditionally favored trips: the "seven carries" trip and the "nine carries" trip. Both originate from Little Green Pond near the state fish hatchery off Route 30 at the head of Upper Saranac Lake. Significantly, the first two ponds encountered—Little Green and Little Clear—harbor brood stock for the hatchery; thus no fishing is allowed. However, St. Regis Pond lies just a half mile portage northwest of Little Clear, and this 340-acre pond provides spectacular scenery and wonderfully productive trout fishing.

The two "carries" divide here. To the east, the easier "seven carries" trip is roughly 9 miles long, with five short portages of only several hundred yards apiece. Along the way, Little Long Pond and Bear Pond merit special effort, since both shelter large and quite pugnacious brook trout.

The rougher "nine carries" trip begins at the St. Regis's northwestern bay and proceeds across a marked carry trail to Ochre Pond, and thence to a string of ponds known as the Fish Chain, including Clamshell, Fish, Little Fish, Lydia, Nellie, and Bessie Ponds. One of these ponds is more beautiful than the next, and leaving is never easy. However, there is the benefit of an alternate route, which takes you through Turtle and Slang Ponds to takeout points at Hoel or Long Pond.

The sight that greets anglers as they approach these lovely wilderness ponds is often hundreds of fish dimpling the surface as they feed on emerging insects. In many ponds, you can be virtually certain that every fish is a brook trout, courtesy of the DEC's extraordinary pond reclamation program in which competing fish (e.g., golden shiners, perch, etc.) are eliminated and heritage strains of brook trout unique to the Adirondacks are stocked. (In some ponds, more domestic strains of trout are stocked, and where depths permit, even lake trout.)

The size of these trout can be stunning. Senior Aquatic Biologist Richard J. Preall, who takes considerable pride in the DEC's work in the St. Regis Area, cites Black Pond as a perfect example of how well trout can do when competition is eliminated. "We reclaimed Black Pond in the fall of 1997 and stocked it with 4-inch native strain brook trout in the spring of 1998," Mr. Preall told us. "By October they were 14 inches, having gorged on crayfish and invertebrates that proliferated when other fish were eliminated." He added, "I've seen 2-year-old brook trout 20 inches in length."

Tip: Anglers are cautioned that no live fish may be used as bait in any of the ponds, since many have been reclaimed for trout. A wide variety of artificials produce action, especially fluttering spoons and leech imitations trolled at varying speeds. As its name implies, the St. Regis Area is canoes only, with no motors of any kind permitted.

The St. Regis Canoe-Only Area offers pleasures that might be astounding to those from other parts of the country more generally acknowledged for their unspoiled natural heritages. Consider the joy of an unspoiled pond shadowed by poplars, spruce and hemlock, where the only noise other than a loon's call is the sound of your paddle and the vista ahead is a surface dimpled by a hundred trout feeding at sunset.

48 Saranac Lakes

Directions: Take Route 86 to the town of Saranac Lake, then turn onto Route 3 for access to Lower and Middle Saranac Lakes. At the foot of Upper Saranac Lake, turn onto Route 30, which tracks and provides ample access to the upper lake for its entire 7-mile length.

Saranac Lakes might best be viewed as a complex system of interconnected fishing waters, each with its own distinct possibilities. The four most prominent features of this system are the Upper, Middle, and Lower Saranac Lakes, all connected by the fish-rich Saranac River. In addition, the system includes a number of smaller lakes and ponds, including Oseetah Lake, Kiwassa Kale, Lake Flower, Fish Creek Pond, and Lake Colby, each providing its own unique habitats and opportunities.

The largest and deepest of the lakes is the Upper Saranac. More than 7 miles in length and as deep as 100 feet in spots, the upper lake is home to significant populations of lake trout and land-locked salmon—species which are largely absent from the remainder of the Saranc system. Throughout most of the year, both lakers and salmon are found in deeper water, most often by anglers skilled in the use of downriggers or

Figure 15 ■ Lower Saranac Lake is dotted with islands, rocky points, and drop-offs and loaded with bass and northern pike. *(Photo: Saranac Lake Area Chamber of Commerce.)*

deep live bait presentations. However, in early spring, big lakers roll in the shallows and often are taken on spoons, spinners, and other artificials fished near the surface. Lakers in the 5- to 10-pound class are common, and more than a few 20-pounders have been caught.

Middle Saranac Lake and Lower Saranac Lake are essentially warm-water fisheries, best known for bass and big northern pike. The lower lake—the most scenic of the three—is dotted with islands, drop-offs, rocky points, weedbeds, stumps, and logs. Both smallmouth and largemouth bass are found in good numbers, and are taken on the usual array of live and artificial baits. Reminiscent of Deep South bassing, popping bugs on the surface at dawn and dusk in early summer yield especially exciting results here.

Bass anglers casting to weedbeds and marshy areas of the middle and lower lakes often are startled by a bone-jarring strike, a brief surge and then a "broken"—actually, a cut—line. This is the trademark hit of voracious northern pike, which frequent the shallows. Despite the northerns' size, few anglers fish the lakes specifically for them—that is, deliberately work the weedbeds and bays with wire leaders. Most are taken by bass anglers who hook the fish in the side of the jaw where the monofilament does not pass over razor-sharp teeth.

Lake Colby, situated at the foot of the Lower Saranac, is notable for the great variety of its fish and the seasons it presents to anglers. This smaller lake is the most consistent producer of brown and rainbow trout. In addition, the lake produces bass, pike, splake in large numbers, big yellow perch, smelt, and bullheads. Most surprising, perhaps, the lake provides a thriving kokanee salmon fishery, especially for the large numbers of hardy souls who fish in the dead of winter when the lake ices over.

Tip: Try trolling for kokanee offshore in the fall, not too far down, with a spoon. Put one or two corn kernels on each hook and dab a drop of fish attractant on each kernel. We caught a ton of kokanee this way in New Mexico's Heron Lake and there is no reason to suspect it won't work in New York.

The lakes are tied together by the main stem of the Saranac River, an outstanding fishery in its own right. Flowing east from headwaters in the High Peaks region of Adirondack Park, the river is one of the few to have benefited by hydroelectric development. The best evidence is the superb walleye and smallmouth bass fishing that has developed behind some of its dams, and the outstanding brown trout populations found in the tail waters below the dams. The North and South Branches of the Saranac are classic trout waters, and the impounded lakes and ponds along their flows help to create fascinating mixtures of cold-, cool-, and warm-water fishing. The North Branch feeds Lake Kushaqua and flows along the southern tip of Loon Lake. The South Branch is impounded to create Franklin Falls Reservoir and Union Falls Reservoir. The river's pools and riffles above these impoundments are classic trout waters; the waters in the impounded lakes are home to significant populations of smallmouth and largemouth bass, northern pike, and walleye.

The region often is referred to as the Tri-Lakes area. However, this does not begin to adequately describe a system of lakes, ponds, and a fine river, which together provide all of the fishing variety and pleasure most anglers might search a dozen locations and hundreds of miles to find. Indeed, as Richard Preall points out, there are three other lakes downstream of Lower Saranac Lake—Lake Flower, Oseetah Lake, and Kiwassa Lake—which themselves offer sensational fishing opportunities. "A day on any one of these lakes and most particularly Oseetah Lake," he notes, "can easily produce a dozen northern pike up to 25 inches long with a good chance that the next strike will be a 10-pounder. And the fishing for big largemouth bass and yellow perch is excellent as well."

Significantly, the DEC maintains scientific stocking programs throughout the Saranac Lakes system calculated to keep the fishing excellent well into the future. In a recent stocking season, for example, the DEC stocked more than 177,000 rainbow, brown, and steelhead trout and land-locked salmon at 24 separate river sites in Clinton, Essex, and Franklin Counties. In addition, the DEC stocked the lakes along its flow, including generous stockings of brown, rainbow, and lake trout in Upper Saranac Lake, and stockings of trout and land-locked salmon (the latter including a few fish as large as 30 inches) in Lake Colby.

To add to the pleasure of the Saranac Lakes experience, good boat ramps are conveniently located at both ends of the upper lake and at the center point near Fish Creek Pond. Additional launch sites are found on Lake Colby just off Route 86, on the Middle Saranac at South Creek, on the Lower Saranac off Route 3, and on Lake Flower in the village of Saranac Lake.

49 Tupper Lake

Directions: Take Route 3 in Franklin County to its juncture with Route 30 at the village of Tupper Lake. Route 30 crosses a portion of the lake and traces its eastern shoreline.

Your boat is the first on the water as you push off into a light mist at daybreak. In the distance, a heavy fish rolls in the shallows and then another, and now you are confronted with a typically delicious Tupper Lake dilemma. You had made up your mind to fish live bait in the deeper water beyond Norway Island for lake trout and you are rigged with sliding sinkers and #6 hooks. But the bass and pike are working the weedbeds and rocky points and the opportunity to cast a spinner bait or a shallow-rigged minnow is so tempting.

That is Tupper Lake in a nutshell. Fed by both the Raquette River and the Bog River, the fish-rich 6,227-acre lake is essentially three bodies of water in one. A shallow delta and marsh connect Tupper Lake with Simon Pond and Raquette Pond, which figuratively form the left and right arms of a "Y." Acres of shallow, weedy shorelines and great expanses of rocky bottom provide excellent habitat for bass and pike, with smaller numbers of walleye found along the drop-offs.

Characteristically, the lake's plentiful smallmouth bass are found in 5 to 15 feet of water, where they feed on the lake's crayfish and forage fish. Bronzebacks especially favor the lake's submerged and exposed rocky islands that are surrounded by shallow water.

Northern pike prefer the weedy shallows, where they spawn just after ice-out, and where they wait in the concealment of vegetation or stumps to ambush their prey. The lake's abundant populations of yellow perch, sunfish, suckers, and small bass provide the northerns with steady food supplies. The South Bay section by the Bog River Falls is a particular hot spot for pike because the water flowing over the falls forces a lot of forage fish into the area. Not surprisingly, big northerns gather for easy pickings.

Walleyes averaging 2 pounds but running up to 6 pounds or more are plentiful and are found in any of the lake's inlets and bays fed by feeder streams. Larger walleyes are generally caught by trolling, especially along drop-offs along the channel running directly through the center of the lake.

Tip: Many large walleyes feed well in very shallow water in the deep dark of night. Sportswriter George Skinner provides this tip for successful night-time trolling on Tupper Lake: "Using a loop knot or improved clinch knot, tie a 'minnow type' lure directly onto your monofilament line. Place a medium-size split shot about 2 feet from the lure and troll only fast enough to give your lure the proper action. Troll as close to the shoreline as possible, without running onto the rocks."

Although generally regarded as a shallow-water fishery, Tupper Lake actually has spots running to 100 feet or more in depth—especially in the area between Black Point and Norway Island. These depths harbor first-rate populations of lake trout and land-locked salmon, both courtesy of the DEC. Significant numbers of lakers and salmon are regularly stocked in the lake at Altamont in Franklin County.

The lake provides many splendid vistas, including the sheer 100-foot cliff of Devil's Pulpit on Bluff Island and the 30-foot Bog River Falls where the Bog River cascades into the lake. These features only add to the awesome beauty of Tupper Lake. Moreover, both Bog River and Raquette River provide fine fishing for the angler who loves moving water. Featuring many deep pools, rapids and backwaters, Bog River compares favorably with better-known trout streams in the north country. And bass, walleyes, northern pike, and panfish are plentiful along most stretches of the beautiful Raquette.

Boat access to the lake is via a hard-surface launching ramp maintained by the DEC in the hamlet of Moody, 2 miles south of the village of Tupper Lake on Route 30. In addition, a state boat-launching site is maintained on the Raquette River 2.5 miles east of the village on Routes 3 and 30.

50 Schroon Lake

Directions: Take Route I-87 in Warren County to Route 9 at Pottersville. Route 9 closely parallels the lake's western shoreline.

Schroon Lake is a 9-mile-long glacial lake that brings together all of the most desirable aspects of still water. It is crystal clear, set in mountain scenery as beautiful as an early season dream, filled with game fish and redolent with deep holes, islands, rocky drop-offs, and weedy structure that guarantee superb fishing variety. More to the point, it is a fabled two-story fishery, with depth and cold springs providing first-rate lake trout and land-locked salmon habitat, plus fine populations of largemouth and smalllmouth bass, northern pike, pickerel, and panfish thriving in the shallower waters.

Running to a maximum depth of 150 feet, the lake is best known for land-locked (Atlantic, non-sea-run) salmon, which gorge on smelt in shallow water when winter turns to spring. This is the most exciting time to go for this spectacular fighting fish, which generally treats the angler to several leaps completely out of the water after being hooked. Trolling or casting lures that imitate smelt is the most successful method in the spring.

Once water temperatures reach the mid-50s, the salmon move offshore and into deeper portions of the lake and become somewhat more difficult to locate. However, they are active predators throughout the summer, and skilled anglers utilizing downriggers or weighted lines do well sweep-trolling the thermocline, generally at depths of about 100 feet. In the fall, when their instincts concentrate on spawning, the fish feed less eagerly; however, the occasional salmon will be enticed to hit bait or an especially well-placed lure.

Following similar feeding patterns, husky lake trout often are taken by anglers intent on land-locked salmon. Many trout are taken in shallow water during cool months of spring and fall, and in summer are found lurking in the same deep holes, but generally at greater depths than salmon. Slow trolling along the bottom is recommended, following temperature breaks. One of the most highly regarded hot spots for big lake trout in summer is the channel running north and south from Eagle Point.

The angler intent on a trophy land-locked salmon or lake trout may want to head for the lake in winter to fish through the ice. To some, ice fishing is what an angler does when he or she "gets fed up with being comfortable or sane." For others, however, ice fishing produces its own special pleasures, not the least of which is relief from "cabin fever" that may grip us in the dead of winter.

Tip: Live smelt are the preferred bait through the ice, but shiners or suckers may also produce good catches if smelt are unavailable at local tackle shops. Try suspending bait at different levels below the ice, starting at about 10 feet and working 2 to 5 feet deeper on each successive tip-up.

The lake's abundant bass populations provide their own special brands of excitement for Schroon Lake anglers. Feisty smallmouths are found along the steep, rocky structure on the east side of the lake from Adirondack Village south, and along rocky outcroppings at any of the islands that dot the lake. Largemouths, while less abundant than their bronzebacked cousins, are taken along the weedlines in shallow water on the southern and western expanses of the lake. Lunker northern pike are found in this same habitat, and are susceptible to the same lures and offerings of live bait.

Access to the lake is excellent. At the northern end of the lake, Dock Street and Fowler Avenue provide good approach to especially productive ice-fishing spots. Hayes Road off Route 9 provides good access to the central section of the lake, and the southern basin is reached from the village of Adirondack, where Mill Brook enters the lake. Two boat-launch facilities are provided by the DEC. The first, a hard-surface ramp with parking for forty-nine cars and trailers, is located at Horicon off Route 9, just north of Pottersville. The second, which provides for hand-launching only, has parking for only four cars.

51 Lake George

Directions: Follow Route 9N along the entire western shoreline of the lake, from Lake George at the southern tip of the south basin up to Ticonderoga at the northern end of the North Basin.

To appreciate Lake George is first to understand its spectacular beauty. Known by some as the "Queen of American Lakes," one of many struck by its awesome splendor was Thomas Jefferson, who wrote to his daughter in 1791 that "Lake George is without comparison the most beautiful water I ever saw." He added, "An abundance of trout, salmon trout, bass and other fish with which it is stored have added to our other amusements, the sport of taking them."

From Jefferson's time to today, these words are perfectly descriptive. Lake George remains one of the most beautiful bodies of water in America, and it is still chock-full of large "trout, salmon trout, bass and other fish" providing wonderful sport for anglers at the dawn of the twenty-first century.

Lake George generally is regarded in two parts: the South Basin and the North Basin. The lake is one of New York State's great two-tier fisheries, with fine lake trout and land-locked salmon as well as excellent bass, northern pike, and panfish waters. Both the North and South Basins provide fine land-locked salmon fishing. Although fish generally average from 3 to 4 pounds, trophy fish weighing upwards of 9 pounds are taken, with the occasional 10-pounder coming to net. Lake trout grow larger. Trophy fish in the 15- to 18-pound class have been taken in both basins of the lake. Smallmouth bass are the reigning monarchs of the warm-water fishery, with bronzebacks of 4 to 5 pounds regularly caught around rocky outcroppings and islands that dot the lake. Largemouths in the 5- to 7-pound class are numerous enough to attract bass tournaments each year. Northern pike from 5 to 20 pounds lurk in the shallower, weedier bays, crushing plugs and live bait often intended to entice largemouths out of the structure.

The best fishing in the lake requires careful attention to the calendar and temperatures. The best fishing for lake trout and land-locked salmon is in the spring, starting with ice-out and remaining good well into June. Early season salmon are taken around the mouths of smelt-spawning streams. In the South Basin, trolling from Diamond Point down to Tea Island and from the north end of Long Island to the brook in Warner Bay and across Northwest Bay from Indian Brook along the drop-off to Tongue Mt. Point often produces action in depths from just below the surface down to 90 feet. Anglers pick up smallmouth bass at hundreds of rocky points, with special action found along shoreline spots at Tea

Island, Diamond Island, Canoe Island, and the islands found in the narrows at about the lake's midpoint. Largemouth bass and northern pike share shallow waters and weedy points off of Dunham's Bay, Huddle Bay, Basin Bay, Andrews Bay, and Northwest Bay in the south basin. As in other large lakes, anglers fishing the shade pockets and breaks score best on bright days.

In the North Basin, the hottest action on land-locked salmon and lake trout is also during the smelt run, beginning at dawn. Trollers who score big work back and forth across the 200 yards above and below the mouths of smelt-spawning streams. (Work the same areas, but deeper, as daylight hours increase the light.) Later in the spring, special hot spots for both salmon and trout include trolling paths between Hague Point north to Friends Point, the structure along the shoreline from Glenburnie Brook and Gull Bay Brook, and along the island between Huletts Landing and the Narrows. Smallmouth action is traditionally excellent in the rocky structures of the islands off Hague, around Friends Point, and the shoreline north and south of Gull Bay and Prisoner Island. The hottest largemouth bass and northern pike fishing is in the shallow water at Mossy Point, Dark Bay, Indian Bay, and Sunset Bay.

Fine populations of pickerel, sunfish, yellow perch, and crappies are found in the same habitats as smallmouths and largemouths.

Tip: It is crucial to remember that smelt cannot be used as bait on Lake George under any circumstances. Most land-locked salmon and lake trout are taken on artificial lures, crankbaits, and spoons, the latter to include Krocodile in blue/nickel and pale green and Mooneye in silver, gold, and copper. (Favorite crankbaits include the dependable Rapala and Rebel lures in a variety of colors.) However, the angler interested in live baits can often use emerald shiners and worm trailers on Lake Clear Wobblers to good effect. Mooselook Wobblers are extremely popular for land-locked salmon.

The usual array of natural baits are fatally attractive to both smallmouth and largemouth bass, especially crayfish, minnows, emerald shiners, night crawlers, worms, leeches, and small frogs. Bass also are regularly taken on a wide variety of plastic worms, spinner baits, crankbaits, spoons, jigs, and surface plugs. Northern pike, always voracious fish eaters, show a marked preference for large minnows, shiners, and suckers up to a foot in length. It is crucial to match your hook to the size of your bait. (We both remember a conversation about the size of hook that should best be employed when using a big yellow perch as bait for northerns. RB suggested a larger hook; ML said his hook would do fine, thank you. Moments later, a massive northern put RB in a position to say, "I told you so," an admonition he had the good grace—or was it good sense?—to stifle.)

Active catch-and-release programs in Lake George and other Warren County fisheries encourage anglers to release large lake trout, bass, northern pike, and land-locked salmon. If a large fish is photographed and released, the angler receives a Warren County "Catch/Release Patch," a certificate, and a membership card from Warren County.

Bob Brandt, a supervising aquatic biologist with the DEC and an authority on fishing Lake George, is especially enthusiastic about lake trout fishing through the ice, using live suckers at bottom. The lake has no closed season for lakers, and current regulations allow two of 23 inches or more to be taken home. (The limit on land-locked salmon also currently is two, with minimum length of 18 inches.) Bob's minimum thickness for safe ice is at least 5 inches. A good rule of thumb, here as in other lakes, is never to be the first person venturing out onto the ice.

There is a lake usage fee for boats on Lake George, administered by the Lake George Park Commission. These fees are used to maintain water purity and to provide a safe and pleasant experience for boaters.

Access to the lake is excellent, with fifteen commercial boat-launch sites on the west side of the Southern Basin (at Lake George Village, Diamond Point, and Bolton Landing) and another five launch sites on the east side of the basin. Boat launching also is available in the North Basin at Lake Shore Drive in Silver Bay. The DEC maintains launch sites at Rogers Rock Campground near Hague, at Northwest Bay Brook north of Bolton Landing, at Mossy Point 2 miles north of Ticonderoga, and at Million Dollar

Beach in the village of Lake George. More than twenty boat rental facilities are located along the North and South Basins.

52 Oneida Lake

Directions: Take U.S. Route 81 through Onondaga County to Route 37 at the toe of the lake. Follow Route 37 to Route 49, which traces the upper end of the lake in Oswego and Oneida Counties.

Oneida Lake is the largest inland lake in the state, covering more than 51,000 acres, and it has been said to offer anglers more fish per acre than any other lake in the northeastern United States. Fifty-eight species of fish grow in this extraordinary fishery, including perch, smallmouth and largemouth bass, pickerel, northern pike, and big channel catfish. However, when most anglers dream about returning to Oneida Lake, it is walleyes that fire their imaginations.

Certainly Oneida Lake's greatest claim to fame is its walleye fishery. Most authorities believe the lake provides the finest walleye fishing in the state, superior even to Scholarie Reservoir. Although populations vary significantly from year to year based on environmental conditions and abundance of forage fish, Oneida has more than a million adult walleyes in successful seasons.

The principal prey of walleyes is the yellow perch, whose numbers in the lake may fluctuate from one million to six million in any given year. Not surprisingly, walleye are easier to catch in years when yellow perch numbers are trending downward and the "eyes" are hungry. Further to appetites, walleye consume six to eight times their weight in prey each year, including their own when such species as young white perch, gizzard shad, and yellow perch are scarce.

Spring is the hottest time for walleye fishing. By the middle of May, the "eyes" become very active, with fish averaging 16 to 18 inches, and the occasional trophy of 5 to 8 pounds coming to net. Feeding mainly from late evening through black dark, and showing a preference for cold, blustery weather, walleyes are essentially bottom-feeders. Consequently boat fishing is preferable to shore fishing. Successful trollers work lures deep and bait anglers drift or anchor along the deeper parts of the lake.

The warm months of summer provide prime fishing conditions and fine catches of smallmouth bass, largemouth bass, and big perch. Characteristically, the smallmouth bass congregate around structure, including drop-offs, rock piles, and rubble, while their largemouth cousins are found at the edges of breaks in the weeds. Yellow perch are found throughout the lake, from the shallow weedy areas around Frenchman's Island to deeper water at the dam.

In the late fall, walleyes once again dominate, especially in the deep drop-offs on the lake's southern side. Hot spots for walleye include drop-offs at Willard Island, Big Bay, Cleveland Bar, and deeper water just west of Frenchman's Island. Walleye often travel in schools, much like the yellow perch they stalk. Consequently the skilled walleye angler gets his or her line back in the water very quickly after a fish is boated.

When the lake freezes, hardy ice anglers dot the lake in quest of walleye and yellow perch.

Tip: Best fishing times through the ice at Lake Oneida are pretty much the same as spring and fall for walleyes, dawn and dusk. For best action, start early so that you're set for action at first light. Remember that lively bait produces the most action; consequently killies, suckers, and chub are most productive, since they never stop moving. And don't forget your spinning rod. Jigging effectively outproduces downriggers at least two to one if you have the patience and the skill to work jigs properly.

The DEC contributes greatly to the future success of walleye angling in Oneida Lake, setting angling regulations year to year based on projections of walleye abundance. Minimum length limits are lowered to increase the harvest of surplus walleye when stocks are high and raised to reduce harvests when the population drops. Moreover, the Oneida Hatchery at Constantia takes eggs from thousands of females netted in the lake and stocks 100 million fry. (Interestingly, between 4 and 8 billion eggs are naturally produced by walleye in the lake each year, but only about one in a thousand of these survives to the fry stage, whereas well over half of the eggs hatched in the nursery become fry. Thus In most years, well over half of the fry in the lake are of hatchery origin.)

Access to the lake is excellent, including boat launches for the large motor craft that are allowed on the lake. The best boat-launch facility can be found off Route 31 in Madison County, 1 mile east of the hamlet of Bridgeport, providing four concrete launch ramps and parking for 100 cars. Two other ramps, also are maintained by the DEC: the Godfrey Point launch, located adjacent to Route 49 just east of Cleveland in Oneida County, and Three Mile Bay launch, also off Route 49, 5 miles southeast of the village of Central Square. Additional shore access is provided at the Big Bay State Wildlife Management Area and the Three Mile Bay State Wildlife Management Area.

53 Great Sacandaga Lake

Directions: Take the New York State Thruway to N.Y. Route 30 in Fulton County. Follow Route 30 to points where it touches the lake, including Cranberry Creek and Northville.

Great Sacandaga Lake once was known simply as Sacandaga Reservoir. Some suggest its name was changed to reflect the world-record northern pike caught in 1940. The 51.5-inch monster, which tipped the scales at 46 pounds 2 ounces, stood for almost four decades as a world record and remains the North American record to this day. Or perhaps it is the mammoth size of the lake that the new name was meant to impart. At 29 miles long, Great Sacandaga Lake lies in three counties (Hamilton, Fulton, and Saratoga) and is longer than some rivers. In our view, however, the "great" was most properly added to reflect the fishing in a lake that still yields big northerns, more than a few 4- and 5-pound smallmouths, brown trout in ever-growing sizes and numbers, and impressive stringers of walleyes and panfish.

The lake is fed by a number of streams, many of which are stocked with trout by the DEC. Many of these trout make their way into the big lake, which is itself stocked with browns and rainbows. In a recent stocking season, for example, the Sacandaga River was stocked with thousands of brown trout ranging from 8 to 14 inches in eight separate locations in Hamilton County. Sand Creek, Hans Creek, and other streams that feed the Great Sacandaga also are stocked with trout. These stockings support a growing brown trout fishery in the lake where browns as large as 5 pounds are being taken in spring and early summer. Ice fishing is popular here, with local anglers scoring well on trout, pike, and yellow perch.

The lake's abundant smallmouths are especially active in early morning and as shadows lengthen in evening during springtime, feeding in shallow bays and around structure in coves. However, as the days grow longer, they retreat to holes, undercut banks, or fairly deep water, essentially to avoid bright daylight. Anglers working shiners, hellgrammites, crayfish, and night crawlers deep along the edges of drop-offs score well, including the occasional 5-pound bronzeback.

Sensitive to bright light, walleyes feed in shallower water only in low light conditions, especially very early morning and late evening into dark of night. In summer, Great Sacandaga walleye are found along drop-off edges of large mid-lake points and humps above the summer thermocline. When fish are tight to the bottom or into weed cover, an effective means of inducing them to hit is working worms or jig combinations (with minnow or plastic tails) along the drop-offs and deep weedbeds, letting the jig fall between stalks to trigger fish. Night fishing with diving plugs can be quite efficient, as can working slider floats (slip bobbers) with live bait as the "eyes" work their way into shallower water.

Tip: *As fishing pro Babe Winkelman points out, "Fishing night crawlers on live-bait rigs for walleyes can be a real test for the angler's sense of feel, to say nothing of his patience. After reeling in a chewed-up crawler, most anglers respond by peeling off more line and giving the next fish more time to swallow the bait. While that works once in a while, the best way to know when to set the hook is by feel. Pick up the line in your index finger. If you feel a solid pull, chances are the fish has the bait and you can set the hook. If you feel a 'nibbling' sensation from the other end, quickly drop the rod and play out a little more line. You'll be surprised how effective this system is."*

When the lake freezes over, the walleye fishing doesn't miss a beat, according to Dean Northburg, a resident of Corinth, New York, who cut his teeth on great trout fishing in his native Montana. He takes fine catches of walleye by jigging dorsal-fin-rigged sawbellies through the ice. When warm weather arrives, he switches to Lindsey-rigged leeches, which he works along the edges of sandbars.

Slider floats also are useful in quest of the lake's legendary northerns, most of which are taken from shore rather than by the more conventional means of drifting or trolling. Of course, many pike in the 3- to 7-pound class are taken from boats, which provide the advantage of covering large amounts of territory while drifting or trolling live bait, plugs, or spoons. However, the really big fish tend to be taken from shore by anglers who set up along promising areas and cast exceptionally large live bait rigged on large hooks and light wire leaders. The bait—usually suckers in the 12- to 15-inch class—would be bragging size for steam trout anglers. In this instance, the old axiom about "big bait, big fish" holds extremely true. The best location is along the shallower basin in the southwestern section of the lake, especially around spawning time in late spring and early summer when big pike are still in the shallower sections of the lake. Then the monster "water wolves" lie in wait in the weedbeds. Of course, there is no guarantee that another world-record pike swims in the Great Sacandaga, but certainly northerns as large as 35 to 40 pounds are present in sufficient numbers to keep hopes for a new record high.

Four free state-operated boat launches are available on the Great Sacandaga, near the towns of Day and Broadalbin, at Northville (on the Sacandaga River), and at the Northampton Beach campsite just south of Northville. A number of commercial launches also are located along the 125-mile shoreline of this great lake.

54 Saratoga Lake

Directions: Take Route 9P out of Saratoga Springs, proceeding a short distance southeastward to the lake. Follow 9P across the upper neck of the lake and down its eastern shoreline.

Saratoga Lake is a jewel of a lake that provides nonstop warm-water/cool-water fishing action twelve months a year. Shaped in a roughly oval form, the lake features broad, shallow bays in its southern end, plenty of steep drop-offs in selected spots at both ends of the lake, and depths of up to 90 feet in the bulge of its northern expanse. Along with these variations in depth, the lake has plenty of rocky structure, expanses of gravel and mud bottom, weed growth in the shallows—in short, every kind of habitat productive of cool- and warm-water species.

Perhaps the best fishing in the lake is for walleyes, a phenomenon the DEC is committed to preserving. In a recent stocking season, for example, the DEC stocked 8,678,000 walleye fry in the lake. Significantly, the lake also has natural reproduction of walleye that probably produces more adult walleye than the state's stocking program. Through most of the year, the best walleye action is in deeper water near the southern end of the lake. Big walleye and exceptionally large yellow perch are taken through the ice in this outstanding ice-fishing lake.

Tip: For hot yellow perch action through the ice, make the effort to find water in the 25- to 40-foot-deep range, out from the south or southeastern side of the lake toward the middle. Schools of big, tasty yellow perch will be found near the bottom in such depths, and will respond well to lively shiners. And if you can stand the cold, consider fishing at night if you are certain that the ice is absolutely sound. Hardy anglers score big catches of walleyes as well as perch after dark.

A special hot spot for both walleye and yellow perch is the north end of the lake, where Saratoga empties into Fish Creek. Dean Northburg, who also fishes the Great Sacandaga and the Salmon River, told us that Saratoga Lake is a particular favorite for "eyes," and that the lake's outflow point produces some of the largest perch in the area.

The lake's reputation for large fish also extends to northern pike; indeed, the northerns here are larger and feistier than those in most area waters. When the ice gives way to warmer weather, there is a heavy spawning run at the inlet as big northerns and walleyes make their way toward moving water. The fishing activity is correspondingly heavy once the season opens.

In later spring and summer, the bass season starts in earnest, with big largemouths found in numbers in the weedbeds and around woody structure, and 2- to 3-pound smallmouths found in a wide variety of rocky structures. Large minnows fished at the edge of weeds and woods work wonders on hungry bass and the full array of spinnerbaits, crankbaits, and artificial worms also put bass on your scorecard, right through the chilling winds of autumn.

Finally, we should note that those large fish rolling in the shallows in spring—the ones that set beginning bass anglers racing across the lake with lures at the ready—are not bass at all. They are carp, often huge carp, and if approached with the proper reverence and the right bait and just the right presentation, they can provide long, powerful fights and endless angling pleasures. Izaak Walton had it right in 1653 when he called carp "the queen of rivers, a stately, a good and a very subtle fish." Carp often require a gradual introduction to new foods and may not take what we consider preferred morsels the first time they are offered. However, these powerful fish will generally take small night crawlers in very early spring, and either prepared carp baits or corn kernels through the summer and early fall. Although some anglers have luck with floating baits, our experience favors bait fished on bottom, with line running freely through a small sliding sinker stopped with a small split shot and one's drag left entirely open against the sudden, powerful run that characterizes carp bites.

Boat access to the lake is via a hard-surface ramp maintained by the State Parks Department adjacent to Route 9P at the north end of the lake. Parking is provided for forty cars and trailers.

55 Chautauqua Lake

Directions: Take U.S. Route 20 along the Chautauqua County shoreline of Lake Erie to N.Y. Route 394 at Westfield; follow 384 to the lake and along its entire western shoreline.

"Wall-to-wall walleyes and lots of muskies at the southeastern end of the lake" was the way Patrick Festa, supervisor of inland fisheries,

DEC, characterized Chautauqua Lake, and his description could hardly be more apt. This exceptional lake, which is 17.5 miles long and has a surface area of 13,156 acres, has long been recognized as a premier muskellunge fishery, yielding many fish in the 45- to 50-inch size. In addition, large numbers of oversized walleyes are caught in the lake in all weathers, from cool spring mornings to hot summer evenings to dead of winter by anglers who know how to adjust their locations and presentations.

To add to this cornucopia of fishing pleasure, Chautauqua is one of the finest bass lakes in New York State. Long expanses of shallow, weedy areas yield "bucketmouths" in surprising sizes and numbers, and the numerous rocky points, rock faces, and deep drop-offs are perfect habitat for scrappy smallmouths. In addition, swarms of crappie, yellow perch, and bluegills work the weedbeds, making this lake an ideal place to take the kids for a day of fishing fun.

Chautauqua is virtually two lakes in one, divided by a pinched land formation at its center. The North Basin is deeper, averaging 25 feet but with holes as deep as 75 feet. The shallower, weedier South Basin averages 11 feet and goes no deeper than 19 feet.

Not surprisingly, the angler intent on muskies is best served to work the deeper waters of the North Basin, although large fish are taken along the weedlines and structure of the South Basin as well, especially around Grass Island and Lakewood Bar. Hot spots in the North Basin include the deeper water around Prendergast Point, Dewittville Bay, the flats near Mayville, and Irwin's Bay. Most muskies are taken on artificials, trolled or cast, which imitate injured baitfish. When water clarity is poor, larger, flashier lures tend to trigger more strikes.

Stocking programs make clear the DEC's plans to maintain Chautauqua Lake as a great muskellunge fishery. In 1998, for example, the DEC stocked 10,000 fingerlings plus tens of thousands more of spring fry. Of course, surviving muskies grow rapidly in a lake as filled with gizzard shad and other forage fish as Chautauqua.

The lake's walleyes especially require adjustments in presentation as the seasons progress, but all seasons are good seasons for this exciting and tasty fish. The lake's gravel shoals provide fine catches in early spring, but as the sun climbs in summer, the "eyes" seek the relative shade and cooler temperatures of deeper water. Experienced anglers fish deep during the day, then move to shallow drop-offs at dusk when the fish swim inshore to feed. Especially good areas to fish include portions of both basins near the constriction that separates them, including Bemus Bay, Warner Bay, Sunset Bay, and Arnold's Bay.

Tip: In all the excitement about muskies and walleye, don't forget the wonderful bass fishing here. The best method for taking Chautauqua Lake bass is casting to weedlines and structure with spinner baits, plastic worms, jig-n-pigs, and of course, live bait—especially shiners for largemouths and crayfish for bronzebacks. In spring and then again in early fall the crappie fishing also is fabulous, when small shiners and jigs tipped with bait make the most appetizing offerings.

Finally, large catches of walleyes, perch, and other panfish are taken through the ice each winter on live bait and on jigs combined with minnows. The North Basin around Mission Meadows, Long Point, and Bell Tower are particularly productive. Access to the North Basin during winter is available via Prendergast Point Fishing Access Site and Long Point State Park. Access to the South Basin is available via the Bemus Point Fishing Access Site.

Public boat access to the lake is excellent. The DEC maintains hard-surface launching ramps and parking facilities on Route 17J in the village of Chautauqua and at Bemus Point on Route 394 in the hamlet of Bemus Point. The New York State Parks Department operates a concrete ramp off Route 17 on Route 430 at Lake Chautauqua State Park between Bemus Point and Maple Springs.

56 Greenwood Lake

Directions: Take Route I-87 to Route 17A, exiting in the direction of the Sterling Forest Ski Area. Follow 17A to the lake, then turn left on Sterling For-

est Road, which traces the entire eastern shoreline of the lake in New York State.

Greenwood Lake came into being when debris left behind by a retreating glacier clogged an icy river called the Wanque. Early settlers examined the miles-long lake that formed behind this natural dam and dubbed it "Long Pond," the name by which it continued to be known for many years. Just as appropriately, they dubbed the watershed "Beautiful Valley," a name entirely appropriate to its rugged natural splendor. The icy river and the glacier carved this valley between Tuxedo Mountain on the New York shore and Bearfort Mountain on the New Jersey side, both soaring to more than 1,500 feet above the lake's sparkling surface.

Lying directly on the New Jersey–New York border, Greenwood Lake is claimed in part by both states. The lake was enlarged in 1768 when a colonial industrialist built a 200-foot-wide dam across the river. Another dam, constructed in 1927 in the interest of flood control, increased the size of the lake again to its current 6-mile length.

Like other mature lakes, Greenwood Lake provides an extraordinary variety of fishing opportunities. Over the years, the lake has been stocked with rainbow, brown, and brook trout. Most recently the DEC has stocked the lake with thousands of tiger muskellunge.

In the 1990s, a great muskellunge fishery developed in the lake. Some anglers fish Greenwood only for muskies, commonly using monster lures and/or enormous live baits.

However, it is worthy of note that both tigers and true-strain muskies often display an appetite for somewhat smaller morsels, conventional wisdom to the contrary notwithstanding. Dedicated angler Joseph Truglio, who lives within easy walking distance of the lake, has hooked more than a few monster muskies when crappie fishing with light line and thin Aberdeen hooks. Recently he fought one massive muskie almost to a standstill before the great fish lunged one last time and straightened the hook. In April 1998, he was again crappie fishing but with slight adjustments to his terminal tackle, including an improved 6-pound test line. The tiger

Figure 16 ▪

Self-described "angling fanatic" Joseph Truglio caught this 15-pound 10-ounce tiger muskie in Greenwood Lake in late April 1998. *(Photo: Melinda Truglio.)*

Figure 17 ▪

White perch seldom grow larger than this fine specimen, taken by Joseph Truglio in Greenwood Lake. *(Photo: Melinda Truglio.)*

muskie that took his bait weighed 15 pounds 10 ounces—a superb fish by any measure other than Mr. Truglio's, who pronounced it small next to the ones that have gotten away.

Tip: *Occasionally a big muskellunge will follow a lure to your boat several times, always turning away at the last moment. If this happens, try plunging your rod tip into the water and quickly weaving the lure through a series of figure-eights. Rather than scaring the fish, this last-ditch maneuver often will trigger a strike.*

Perhaps the greatest fishing interest in the lake centers on largemouth and smallmouth bass. Both bass species grow to exceptional size because of abundant food resources and superb habitat. Scores of streams and creeks winding down the mountain slopes of New York constantly renew food stocks and oxygen in the lake, and two large islands plus the many docks and boathouses around its shore provide shelter and structure for bass.

The lake is relatively deep, with mean depths about 17 feet and many holes as deep as 35 feet. Because of its size, the lake provides a variety of fishing environments, ranging from heavy weedbeds in the shallows to abundant artificial and natural structure at points along the lake's entire 6-mile length.

Some like to recount the baseball legend that hangs over the lake. Babe Ruth, one of many celebrity anglers who fished Greenwood Lake, was on his way into New York City for a World Series game when his car hit a tree on Tuxedo Mountain. The immortal "Sultan of Swat" was reduced to hitchhiking a ride to the Series. (He arrived in time to play nine full innings.)

57 Clove Lakes

Directions: From Verrazano-Narrows Bridge, take the Staten Island Expressway to Clove Road exit to Victory Boulevard and the lakes.

Fishing in the five boroughs of New York may seem a stretch, given the city's reputation for noise, excitement, and concrete. However, the mouth of the Hudson River is one of the richest marine resources in the country.

Breezy Point Jetty in Queens is one of the premier surf fishing spots on the East Coast. Prospect Park in Brooklyn and Van Courtland Park in the Bronx offer fine freshwater bass fishing, and the Connetquot River on Long Island is a storied trout fishery. Central Park's lakes and ponds offer bass, pickerel, carp, bluegills, perch, and catfish for the urban angler. And ranked with the best is Clove Lakes Park on Staten Island.

In fact, Staten Island has a number of quality freshwater fishing spots, including Ipe's Pond, Brady's Pond, and Silver Lake. However, the three ponds of Clove Lakes Park—Clove Lake, Martling's Pond, and Brooks Pond—may be the best of the lot. All three ponds are city-owned and are stocked. Clove Lake is at Clove Road and Victory Boulevard. Clove Road also leads to Martling's, and then Brooks Pond, which is located near the Staten Island Zoo.

The ponds are stocked predominately with largemouth bass, but all three hold crappie, common sunfish, bluegill, carp, and catfish as well. According to Pat Scaglione, proprietor of nearby Scag's Bait and Tackle, Clove Lake has the largest bass, running up to 6 pounds, plus monster catfish. Best baits for bass are live shiners, while the crappie prefer killies. Prepared "Uncle Josh"-style baits are lethal for catfish, and of course, lively night crawlers will catch virtually anything in the ponds, including carp in very early spring.

Tip: While he doesn't sell it, Pat's personal favorite bait for carp is a small floating chunk of a poppyseed bagel. With a breeze at your back helping you cast, chuck out a hunk of bagel on a #6 baitholder hook and watch it float. If you're lucky, a big carp will "inhale" it, and the next thing you know your drag will scream. (Manny can personally attest to a longstanding carp population in Clove Lake. Perhaps 45 years ago, he took the Staten Island Ferry across from Brooklyn with two friends and fished Clove Lake, catching several carp on corn kernels fished at bottom.)

If your angling tastes run to artificials, Pat advises that any color plastic works, just so it is dark. Berkeley Power Worms 4 to 7 inches long work well.

Since New York maintains the park, there is good public access to the ponds. Standard license regulations prevail. Overall, all three ponds are quite beautiful, feature good fishing, and are contiguous to other outdoor pursuits (bike and jogging paths, horseback riding, baseball and soccer fields, basketball courts).

Admittedly, the fishing may not compare to the Great Lakes or the Finger Lakes or the major reservoirs and rivers featured elsewhere in this book. However, the ponds in Clove Lakes Park have features that none of these can boast, including immediate accessibility for countless urban kids to whom a trip on the ferry and a feisty bass or a stringer of bluegills may add up to the memory of a lifetime.

58 Lake Ronkonkoma

Directions: Take the Long Island Expressway to Pond Road at Ronkonkoma. Follow Pond Road a short distance to the lake.

Lake Ronkonkoma is a paradox: a first-rate freshwater fishing lake that is bracketed by the Long Island Sound to the north and the Great South Bay an equal distance to the south. Because Long Island is so famous for saltwater fishing, its excellent freshwater opportunities are often overlooked. In fact, there are 526 ponds on Long Island, ranging in size from less than an acre to 243-acre Lake Ronkonkoma, the largest on the island.

For knowledgeable anglers, Lake Ronkonkoma is rivaled only by the Connetquot and the Nissequoque Rivers for quality freshwater fishing on Long Island. Significantly, the lake has changed in interesting ways over the past two decades. In the 1970s, Ronkonkoma was known essentially as a fine largemouth bass lake, with good fishing for panfish as well. However, the bass fishery declined somewhat, probably owing to a combination of factors including a rapid growth in the white perch population (white perch compete actively with young bass for food), overfishing for bass, and low reproductive success of bass during the period.

In response, the DEC began stocking walleye in the lake in 1994. Walleye are voracious predators with a marked preference for spiny-rayed forage fish such as white perch and yellow perch. Walleyes have demonstrated a significant ability to control overabundant perch populations without adversely affecting bass populations. In fact, walleyes benefit bass populations by reducing food competition for bass fingerlings. This is precisely what is happening in Lake Ronkonkoma, where the walleyes stocked in fall of 1994 have long since reached the 18-inch legal size limit and are cutting a swath through the perch populations. The lake today is a first-rate walleye fishery with good fishing again for largemouth bass as well as pickerel, bullheads, carp, bluegills, and of course white and yellow perch.

Significantly, the stocking program is a continuing process. In 1994, the DEC's Region 1 Freshwater Fisheries Management Unit stocked 850 yearling walleye averaging 8 inches in length. As the walleye presence began to be established, the stocking plan called for 5,000 advanced fingerlings ranging in length from 4 to 5 inches. In a recent stocking season, these 5,000 fish, plus another 5,000 1.5-inch fingerlings, were stocked.

Tip: Of all approaches to catching walleye, combination spinner–live bait rigs have scored best through the years in Lake Ronkonkoma. Vibration and flash from the spinner blade attract a walleye's attention, and the live bait—whether a minnow, night crawler, or leech—induces the fish to strike. Spinner–live bait rigs work best in low-clarity water but will catch walleyes anywhere. Try slowly trolling the lake's large pieces of submerged structure with small silver or brass spinners rigged on monofilament leader with a #6 baitholder hook and small head-hooked night crawler. Fish cooler water offshore in summer but shallower water inshore in the fall.

Lake Ronkonkoma is one of the most accessible bodies of water for its size in the state. Shore access is plentiful. The DEC maintains a hand-launching site for boats with parking on Victory Drive in the town of Islip. (No gasoline motors are permitted.) In addition, a special facility is maintained at the pier in Lake Ronkonkoma County Park for disabled anglers.

59 Fort Pond

Directions: Take Route 27 (Montauk Highway) to Edgemere Road in the village of Montauk.

Start a conversation about Montauk with New York saltwater anglers and they will tell you about charter boats and party boats, surf casting and reef fishing, huge mako and thresher sharks, striped bass, bluefish, codfish, weakfish, and fluke. Mention great freshwater fishing in that same conversation and you'll probably be treated to looks of puzzlement if not derision. However, strange as it may seem, some of the best sweetwater fishing on Long Island is located close to the ocean near Montauk.

The spot is Fort Pond, which has been called "the most dynamic still water on the island." The pond covers 192 acres and runs to maximum depths of 26 feet. It is so close to the ocean that one might expect to take flounder as in Montauk Lake, a freshwater pond until it was opened north into Block Island Sound in the 1920s. However, the primary game fish in Fort Pond are black bass with smallmouths in the 2- to 3-pound range and largemouths often running to 5 pounds or better.

Because of its sweet water and isolation from pollution, Fort Pond is a virtual laboratory for the DEC's stocking concepts. The most exciting to us was the introduction in the early 1980s of hybrid bass, which is unquestionably our favorite freshwater game fish. The hybrid is a cross between striped bass and white bass and combines the best qualities (speed, power, exceptionally fast growth) of both. The strike of a hybrid is unlike that of any other freshwater fish its size. It combines sudden, incredibly fast runs of a bonefish and a sustained fight reflecting all of the dogged determination of a big blue.

The DEC also maintains an active walleye stocking program at Fort Pond. In a recent stocking season, 4,000 fingerlings and another 4,000 "eyes" in the 4.5-inch range were placed in the pond. This program is bearing considerable fruit; indeed, at this writing special regulations mandate a maximum of three walleye 18 inches or longer in the pond, and the good news is that anglers are starting to fill their limits.

Tip: Light level is the single most important influence on walleye behavior. It determines when and where walleyes feed, and where they spend their time when not feeding. Most anglers know that walleyes have light-sensitive eyes and shy away from bright sunlight. However, it is also important to consider the angle at which sunlight strikes the water. When the sun is directly overhead, most of the rays penetrate; however, when the sun is low on the horizon, the rays are mostly reflected. Fish under conditions of low sunlight penetration, especially at dawn or dusk, or at night.

Fort Pond also provides wonderful panfishing, especially for perch, which make up in abundance for what they generally lack in size.

Boats are allowed on the pond, but no gasoline motors. The DEC maintains a hand-launching ramp with parking for twenty cars off Edgemere Road in the village of Montauk.

Saltwater Fishing

The Mighty Atlantic

Unimaginably rich in marine life, the salt waters of New York provide a virtual wonderland of great ocean, surf, and bay fishing for stripers, bluefish, giant tuna, marlin, sharks, blackfish, flounder, cod, and fluke. From Staten Island to Montauk, from Throgs Neck to Orient Bay, the record fish taken in these waters are the stuff anglers' dreams are made of. As of early 1998, these include a 3,450-pound white shark, a 1,071-pound bluefin tuna, a 1,174-pound blue marlin, a 1,087-pound tiger shark, and a 1,080-pound mako. Equally impressive for their species, New York record fish include a 71-pound longfin albacore, a 19-pound 2-ounce blackfish, an 85-pound cod, a 22-pound 7-ounce fluke, a 99-pound 4-ounce wahoo, a 19-pound 2-ounce weakfish, and a 76-pound striped bass.

Whether you dream of catching a half-ton giant tuna, a 250-pound mako shark, a 40-pound striped bass, or just a mess of feisty bluefish or fluke, whether you fish from charter boats or headboats or your own boat, whether you are a surfcaster or a jetty jockey, New York offers the best of saltwater fishing.

There are famous offshore fishing spots like the fabled Block Canyon and the Hudson Canyon, where giants often weighing in excess of 1,000 pounds wait to challenge the skill, strength, and endurance of captains and anglers. Licensed boats and skilled captains and crews usually take anglers out to these special places in the sea, and for all of us, it is difficult to leave the dock without a heart-pounding sense of suspense and excitement at what lies beyond the horizon.

60 Mud Buoy

Directions: Sail about 16 miles southwest from Point Lookout and look for the yellow marker buoy. The loran numbers on top are 26906 and on bottom they read 43655.

The Mud Buoy is about 2.5 miles west of Seventeen Fathoms and lies between Seventeen Fathoms and Sea Bright, New Jersey. The area, which was used for dumping clean dredge spoils and other materials, is named for the buoy that marks it. The dredged materials attract massive numbers of sand eels, rainfish, and squid, all attractive to larger fish.

Boats journey out to the Mud Buoy from a dozen or more ports to get in on the superb bluefishing. In fact, many experienced anglers have come to regard this as the best bluefish grounds of the 1990s. One of these is New York Captain Dennis Kanyuk, who runs his large charter boat, *Princess Marie*, out to the Mud Buoy for blues. He told us recently that although the run is pretty long, the catch is worth the trip more often than not.

The reason is the sheer excitement and fun of catching bluefish, as captured by author John Hersey in his book, *Blues*. "Blues are both butchers and gluttons," Hersey wrote. "They're cannibals that will eat their young. They will eat anything alive. They have stripped the toes from surfers in Florida."

Tip: For bluefish, especially 1- to 3-pounders, use a single hook, without wire leader. Yes, you will be cut off occasionally by the "choppers," but when fishing in a crowd, your offering will look far more natural and your bait easier to take without wire.

The water here is as shallow as 38 feet and averages 50 feet. If you reach 60 feet, you are off the dumpsite and too deep. Daytime is best for bluefish, and the best catches are recorded in August and September. A

Figure 18 ▪ Good-sized bluefish like this one are commonly caught at the Mud Buoy on the *Princess Marie* out of Point Lookout. *(Photo: Captain Dennis Kanyuk.)*

month or so later, sea bass begin to be caught here in numbers. Blackfish also are caught along the area's rock-strewn bottom.

61 B. A. Buoy

Directions: The B. A. Buoy is located 16 miles southeast by south of Atlantic Highlands, New Jersey.

The B. A. Buoy marks the beginning of a vast area famed for rich offshore fishing and is a prized ocean spot to which serious saltwater anglers sail from many New York ports. The area is roughly 16 miles offshore.

Along this trail, too, are the Scotland (to the west) and the Ambrose (to the east) channel buoys. From the buoy south, this area is an enormous shipping lane. (In fact, ships bearing the names *Scotland* and *Ambrose* had been painted red and for many years were anchored as stationary markers for captains to guide their cargo ships.) Fishing in this area is often nothing short of spectacular. In the area of the buoy, in depths averaging 120 feet, ling are a primary target. For much of the year, they are commonly found in numbers on both the soft, open bottom and around the many wrecked boats that lie in the mud.

Big blackfish and sea bass also are caught here in exceptional numbers. Occasionally, cod and pollock are found on the harder bottom sections, especially from midwinter to the middle of spring. Interestingly, once a breakup of Hudson River ice occurs, the seawater is dramatically cooled by huge chunks of ice charging out to sea. Shallower water gets cold more quickly, of course, but even at the B. A., temperatures sometimes drop below the comfort zone of most of these fish. A mild winter can see this area producing action virtually nonstop, but a heavy ice melt often signals the end until April.

Back when whiting were more plentiful, the season began in earnest in November in the Scotland/Ambrose areas, but as the waters cooled, the fish moved to the deeper, warmer waters at B. A. Thus the "winter fishing" for these "frostfish" actually began in March or April around the B. A. Buoy.

Tip: Since you are fishing in deep water, leave your whippy rods home. You can use fairly light line (20- to 30-pound test monofilament) but a 7- to 8-foot boat rod or an extra-stiff bait casting or spinning rod will allow you to set your hook effectively.

Crab bait is best for blackfish, and a strip of herring often produces the best action with most other species, especially whiting and ling. Skimmer clam also works well. Listen to your boat captain for best results.

In the past, tuna were caught in the area of the buoy on a regular basis, especially schoolies weighing 20 to 40 pounds, but occasional giants of 750 pounds or more as well. As recently as mid-December 1997 some anglers were finding that the ling they were reeling up were being gobbled by monster tuna. Something to think about when cruising out to fish the B. A. Buoy.

62 Seventeen Fathoms

Directions: Sail 30 miles from Fire Island Inlet out of Captree on a 250-degree course, traveling west by southwest.

Seventeen Fathoms is an ocean hot spot in the general vicinity of B. A. Buoy and the Mud Hole—frequent destinations for New York anglers. Giants swim the depths of these great fishing areas and wait to challenge anglers' skill and physical abilities. An 880-pound tiger shark, a 1,046-pound blue marlin, a 1,030-pound bluefin tuna, and a 530-pound swordfish are among the record leviathans taken in these waters. Of greater interest to larger numbers of New York anglers, the fishing for blackfish, bluefish, and codfish is often nothing short of sensational.

The eastern edge of Seventeen reaches into the deeper waters of the Mud Hole. The name "Seventeen Fathoms" derives from its depth of just over 100 feet. (A fathom is 6 feet. Dividing 6 into depth here yields the name.) Of course, this depth recommends stiff-action rods for properly setting hooks. Seventeen is a vast area of rocks and snags, all very attractive to marine life. It acts like a magnet to fishing boats from Long Island, Sheepshead Bay, the Atlantic Highlands, and Belmar.

Figure 19 ■ These good-sized blackfish were caught at Seventeen Fathoms on the day after Thanksgiving, 1997. *(Photo: Captain Tony Castaldi.)*

When the water begins to get cold each fall, the area comes alive with blackfish and school-sized cod. Ringers hedge their bets by rigging a blackfish hook or two at bottom, baited with green crab bait, and a size 7/0 codfish hook 5 or 6 feet up the line on a 2-foot leader, baited with a whole skimmer clam, for cod.

Another surefire "tog" and codfish bait is conch. On Election Day, 1951, Manny took the pool on a Sheepshead Bay headboat with a 20-pound cod, taken on tenderized conch meat. It works well to this day.

Tip: Whatever your bait, it is important to check it from time to time, even if fishing well above the rocks for blues but especially if fishing at bottom. The reason: huge schools of talented bait thieves called bergals hide in the snags at bottom, waiting to pick off your bait before a more desirable fish can get to it. (One reason blackfish experts prefer crab or conch baits is that these baits are more difficult to steal.)

Bluefish are caught in large numbers at Seventeen, especially at night. The blues taken here are generally bigger than those taken closer to shore. Big blues, chunky blacks, and schoolie cod make Seventeen more than worth the ride.

63 Cholera Banks

Directions: Head due south, 180 degrees, out of Jones Inlet, for 8 1/3 miles.

The Cholera Banks were famous a half-century ago for incredible codfishing. When the noted fishing writer Ed Hurley began to write about codfishing in the long-defunct *New York Mirror*, readers knew that cold weather had arrived. Simply put, cold weather meant codfishing, and codfishing meant Cholera Banks.

Although the occasional cod still is taken in this wonderful fish-rich area, the codfishing has largely petered out. However, the fishing is often superb to this day, especially for ling, bluefish, false albacore, fluke, and mackerel. Captain Al Lindroth, who runs his boat, *Capt. Al*, out of the Point Lookout Marina at the western corner of Jones Inlet, recently shared his encyclopedic knowledge of the banks with us, beginning with geography.

Figure 20 ■ Tuna.

The estimated depth of the Cholera Banks is 75 feet. Unlike many of the reefs in the area, Cholera is natural rather than artificial, composed of natural rock formations, mostly porous sandstone. No buoy marks the site; however, it is marked by a series of lobster pots in warmer months. Indeed, the pots mark just the right area to start fishing.

The first fishery each year is ling, which appear in April. The warmer the winter, the earlier the appearance of these unattractive but extremely tasty fish. Ling are generally caught at bottom, most on hooks that actually are dusting rocks or sand. Occasionally, however, a three-hook rig will entice a ling to rise above the sinker to take bait.

Tip: The conventional wisdom is that clam baits are the right prescription for ling. However, "ringers" often carry mackerel, since a square chunk of fresh or even frozen mackerel may entice ling to strike more quickly than clam baits.

As the waters of Cholera warm, sea bass join the ling, also feeding on or near bottom. A strip of squid may work, but principally anglers bait with a hunk of clam above the sinker for bass, and clam or mackerel below for ling. Hooks of choice are a 1/0 or 2/0 beak-style Eagle Claw or Mustad with baitholder. Some anglers prefer a rig that makes fishing easier, with a snelled contraption that has two or three hooks, a swivel on top, and a snap swivel below to stop the sinker. Our experience favors the "KISS" (Keep it simple, stupid) method with uncomplicated terminal tackle that always seems to catch more fish.

From July 4, bluefish, bonito, and false albacore are the featured quarry at Cholera Banks. Chumming with ground "bunker" bring these speedy sportfish to baited hooks, but casting and retrieving a 2- to 4-ounce "Ava"-style jig also produces wonderful sport. In fact, jigging is often the preferred style by day with baitfishing preferred at night. (Ask anglers fishing on bottom for any small bergals they catch. Hook a bergal on and let it out in the current. Bigger blues often prefer this natural bait.)

Big fluke also are caught by drifting in and around Cholera Banks. Predictably, anglers hang up in the rocks at bottom; however, the fishing (and fresh fluke fillets for dinner) are more than worth the loss of tackle.

When the waters are warm enough in late fall, Lindroth also takes extremely large blackfish on Cholera Banks, generally on green crab but occasionally on tenderized conch meat. (If your headboat carries a bushel of clams, a conch or two may be mixed in with the skimmers.)

Finally, always have a mackerel rig with you when you fish Cholera. While most of the mackerel have run farther offshore in their migrations over the past few years, a good number appear closer in from time to time. And you will want to be ready for them.

64 Hudson Canyon

Directions: Sail 160 degrees in a south by southeast direction from Fire Island Inlet for 70 miles to the inshore tip of the canyon.

The Hudson Canyon is a fabled tuna-fishing site—perhaps the most avidly fished location for tuna in the northeastern United States. Boats from virtually all of the Long Island and Brooklyn ports make for the canyon, where they meet many more boats from ports in the Garden State.

The banana-shaped canyon is nearly 20 square miles in size and runs from 50 to more than 100 fathoms in depth. The southwest corner of the canyon is referred to as "The Letters"; the southeast corner is called "The Point."

Captain Tony Castaldi, who runs the charter craft *Terminator* out of Captree, fishes the canyon from the beginning of August until early October, with September being his favorite month by far. The *Terminator* is licensed for fifteen passengers inshore and carries as many as ten for a canyon run, owing to its broad beam and large cockpit.

When more than the usual complement of anglers on a standard "six pack" go out for tuna, they work out details in advance regarding the order of fishing when trolling during daylight hours, since not everyone can fish at the same time. It is only at night, when chumming and fishing with bait on anchor, that all hands can fish simultaneously.

Figure 21 ■ These 50-pound class yellowfin tuna were caught at the tip of Hudson Canyon on the *Terminator*.
(Photo: Captain Tony Castaldi.)

Castaldi trolls a Green Machine with a few smaller Daisy Chain lures ahead of it. The larger lure appears to be chasing smaller fish, thus inducing tuna to strike. Trolling along both walls of the canyon works well, and locating temperature breaks and signs of life enables skippers to direct their boats into fish.

At night, boats anchor on the edges of the canyon in about 600 feet of water. The *Terminator* carries two quartz Halogen lights that illuminate the water. This brings bait to the surface and often also brings tuna up to the top after the bait. A yellowfin or longfin (true albacore) charging up into the lights is an incredibly exciting experience.

A steady stream of small butterfish and herring chunks thrown overboard creates a chum line that starts fish feeding.

Tip: Captain Castaldi likes to put a chunk of bait on a small hook and have a long-handled net at the ready. When squid come to feed on the baited hook, he nets them and places them in a live well. While most anglers use the traditional hunk of herring or butterfish as bait, he swears by the use of a live squid. The hook of choice is a 7/0 Mustad number 7699.

A word of caution: Although the Hudson Canyon is a truly great place to fish, only experienced skippers with all of the right gear should make for its beautiful blue waters. Absolute minimum equipment: two well-tuned engines, ship-to-shore radio, and radar, with all of the standard flotation and other safety gear as well.

65 Block Canyon

Directions: Block Canyon is located roughly 65 miles due south of Montauk. The section is about 3 miles long and 1 mile wide and requires the use of a good navigational chart and loran plus a quality depth finder to be certain of location.

When anglers speak of a canyon, they are more correctly speaking of the edge, or drop-off, where the bottom of the ocean stops sloping gently and takes a precipitous drop. In some cases it is an actual canyon, an ancient riverbed with vertical sides like the Hudson Canyon, which is an extension of the Hudson River. The canyon closest to Montauk is Block Canyon, which is also called "Fish Tails." The canyon is extremely deep, reaching depths of 1,000 feet or more.

Fishing the edges of the canyon rather than the depths is the secret to success, and success at Block Canyon is usually measured in terms of tuna. There are certainly marlin cruising the edges—sometimes giants of 1,000 pounds or more—and plenty of dolphins (mahi-mahi). But in summer, way out on the edge of the Gulf Stream, it is tuna you are after, and certainly tuna are present.

Bigeye tuna, which may weigh 200 pounds or more, are the main heavyweights caught along the canyon's edges. Yellowfins, ranging from 15 to 50 pounds, also are found here, along with hard-fighting true albacore. If you're fortunate enough to fish the canyon in calm water, you probably will be rewarded with a sight that few anglers ever get to see: beautiful, blue water with birds screaming and diving as they pick off baitfish chased upward by monster tuna and perhaps even marlin.

Chunking for tuna, with chum to bring them close to the boat, is incredibly exciting. Just imagine a submarinelike torso charging within 10 feet of the gunwale!

Tuna usually bite by day but also feed actively in darkness on bait while anchored. During daylight hours, dolphins provide wonderful entertainment as you troll the edges and temperature breaks. Trolling is almost always the preferred system by day. (Of course, this reduces the number of anglers fishing at any given time on a headboat.)

The very best way to fish the canyon is aboard a charter boat or a headboat licensed for far-offshore sailing. For long outings, an EPIRB (emergency position indicating radio beacon) is required by the Coast Guard. (There are several party boats in Montauk that are so licensed and all have to carry at least two licensed skippers by law. Frankly, we prefer fishing such boats, although you may find more anglers at the rail than you might like at times.)

As noted earlier, if you take your private boat out this far, a number of items are absolutely necessary in the interest of safety. Without doubt, you need a ship-to-shore radio, with a backup radio. You need radar and a good ability to operate it. You need two engines, just in case one quits. You need a high-pitched sounding device and flotation devices, including an inflatable raft, and certainly, good life jackets. To fish open ocean this far from shore, no safety precaution should be overlooked.

Tip: Regarding tackle, forget spinning gear unless you are after dolphin (which are most often found under floating debris and under lobster-pot buoy markers). You might hook a tuna on such tackle, but you'll only create a mess for everyone else on your boat. Fish heavy tackle with the finest reels mounted on good stand-up rods. A fighting chair will help but is not absolutely essential if you have a good rig. Bring a good thermometer to locate water temperature breaks, since the fishing often changes for the better where the temperature changes dramatically.

Several miles east of Block Canyon is the so-called "100 Fathom Curve," where anglers also find excellent action. A typical trip to the canyon involves at least 24 hours, and many outings wrap around two days. Boats sometimes sail late at night, reaching the grounds by daybreak. Big fish are taken at night, but sunup and sunset also are prime fishing times. We recommend you carry a good cooler and plenty of food. Most boats will have some food, but if the action is hot, it may not be enough. Hopefully, you'll need the cooler to transport a fine catch back home. Also, remember to (a) bring a camera and (b) tip the mate. The memories associated with good pictures will warm a cold winter's day when you're stuck indoors, and a hard-working mate earns and deserves a bit of extra consideration.

66 Block Island

Directions: Head due east out of Montauk Harbor for an estimated 17 miles.

Block Island is another fabled Montauk location, one we have fished no fewer than fifty times over many years. Thus reporting this spot is an exercise in wonderful memories of sparkling waters, tight lines, and sometimes frantic action.

A landmark is the Hooter Buoy, located about a mile off the southwest corner of the island. We fished here many times for giant "snowshoe flounders." The waters at the Hooter run from 30 to 50 feet deep and in the spring, especially from mid-May to mid-June, monster "flatties" lead the action.

The time to fish here is at and around slack water. Use a chum pot filled with cracked mussels and clams. Bait with wild mussel, clam bellies, or sandworms as you fish this mixture of sand and mud bottom.

The fluke start to bite here in warmer weather, at about the time the flounder action slows. Fluke can be taken while drifting, especially using a long strip of fluke belly and a sand eel for bait. (Note: no spinners or other "gongs and whistles" are required.) The best action for fluke is found inside and north of the buoy. As you bounce bait along rocky patches of bottom, count on sea bass attacking. (Add a top hook if this occurs. You may hook a bass on the top hook and a fluke on the bottom.)

T.J., of Gone Fishing Marina, adds that striped bass also are now very commonly found around the Hooter. He told us that a wonderful recovery of this fishery has taken place through the 1990s, and that anglers score well on "linesides" by anchoring and baitfishing, drifting with live eels, trolling, and/or jigging.

South of the Hooter is the South West Ledge Buoy, another hot spot for striped bass from mid-May through early December. The bottom is hard in the waters around this buoy, which run from 40 to 70 feet deep. In the autumn, the blackfishing is great here. It is said there are three kinds of bait to use for blacks (also called "togs"). In order, they are green crabs, green crabs, and green crabs.

Tip: If really large "tog" are being caught, stick with the usual two-hook rig, but bait differently. Find a large green crab, break off all the claws, and stick one hook into and out of each main claw opening. Instead of the half crab with top shell removed, use the whole crab without shell. And get ready. If something hits this larger bait, it is usually a beast of a blackfish, perhaps 10 pounds or better.

If you head even farther south from Block Island, you can find depths of 90 to 120 feet about 3 to 4 miles offshore. You will need a good depth finder with loran reading to locate the precise spot, but the fishing may make it well worth the effort. Codfish are found in these waters; not as many as in the glory days of Montauk codfishing, but still enough fish to make the trip potentially very exciting. (Some years back, Manny caught his largest ever in these waters, a 41-pounder.)

Sand sharks also abound here, especially on an easterly wind. Once a mate on a Montauk headboat told us he sold all the "dogfish" he could get from his fares to a famous fish-and-chips (make that "jaws and chips") restaurant. When fishing in this area, try a rig with two hooks, the top one with a 2-foot leader tied about 4 feet above your sinker; the second tied above the sinker knot on an 18-inch leader. If pesky bergals are stealing your bait, try to find a conch in the bait box. Break it out of its shell, tenderize it by hitting it with a hammer, and bait with a big chunk. This will stay on the hook better. Alternately, use a sizeable length of mackerel. In late fall, a bluefish may take this bait but cod like it as well, and it may discourage bait thieves long enough to hook a fine fish.

67 Coxes Ledge

Directions: Run 32 miles straight east from Montauk Harbor. (When you reach Block Island, you are halfway to this huge and highly productive fishing ground.)

Coxes Ledge runs 8 miles long and as much as 4 miles wide in some sections. It is reached after you pass over the second mudhole at what is called "The Gully." The rich waters of Coxes Ledge range from 120 to 140 feet deep and the bottom is lined with rocks. The bigger patches of rocks attract pollack, but the fish of choice here is the codfish, often thought of as the "Winter King." However, some of the experts, prominently to include T.J. at the Gone Fishing Marina, tell us the best codfishing at Coxes Ledge is found in "tee-shirt weather," from June through August.

T.J. also shared his theory on what would appear to be two different species of cod caught in and around Montauk; those that are reddish in hue, and the more typical green fish. He feels the green fish are on the move, and the fish with red skins are from a resident population. He reasons that the resident population lives in kelp-rich waters, which create a natural skin dye. There are other theories, but this one seems logical to us.

Codfish action comes with nearly any kind of natural bait. If the bottom is alive with bergals, try catching and filleting a few. Put a single fillet on your hook, stuck just once, and to that add a whole skimmer clam. If the bait-stealing bergals help themselves to your clams, a fillet of one of their cousins will still be on the hook. Many a cod has been caught on this remnant of bait.

Tip: Instead of the beak hook that many cod anglers prefer, go with a black or brown size 6/0, 7/0, or 8/0 model 3399 Mustad Sproat bend. Snell it with a 40-pound stiff leader that is 2 to 3 feet long. Put just one hook on, tied off of a dropper loop 4 to 5 feet above your sinker. Use a flat sinker, starting with 8 ounces but remain open-minded to as much as 16 ounces, since you are fishing in deep water and often in strong tides. (You may want to tie the sinker on with a breakaway knot in case a big fish pulls the lead into bottom.) Don't use two sinkers, since that spoils your presentation.

Use a stiff-action rod, what we call a "cod rod," one that is anywhere from 7.5 to 8.5 feet long. Unlike a softer rod, this will allow you to haul a big "cow" cod up and out of its rocky home before it hangs you up. Anglers fishing for cod are often surprised in summer by small bluefin tuna. Those specifically targeting the schoolie tuna troll for them, using high-speed lures such as the Green Machine. However, more than one unsuspecting angler reeling up a little codfish have experienced a sudden run-off of line as a tuna of 50 pounds or more grabbed their fish.

Blue sharks are often taken at Coxes Ledge, together with quite a few makos. Manny and fishing buddy Roland Hagon were fishing the ledge for cod when they simultaneously hooked big sharks. Unfortunately, Hagon's 100-pound shark opened its mouth as it crossed Manny's line, neatly slicing through it. The result was one shark caught rather than two. Anglers who are out after sharks will chum for them with ground bunker, mixing in some small hunks to add to the attraction. For bait they use fillets of mackerel (just what Manny and Roland were using for codfish when their simultaneous sharks struck) or bluefish. The hook the "ringers" use is a 10/0 Mustad model 7699 tied on an 8- to 10-foot wire leader.

Wrecks and Reefs

Wrecked ships and artificial reefs provide extraordinary marine habitat where only barren ocean floor previously existed. In many cases this habitat is incidental to a more dramatic history. For example, German U-boats sank a number of tankers and freighters just off the New York coastline during the early days of World War II. Other ships and even an early-warning radar tower have broken up in storms and settled to the bottom. Or in the case of artificial reefs, concrete and steel from demolished structures, thousands of old tires, surplus Army vehicles, and boats that have outlived their usefulness are sunk in often-massive reef areas to provide new places for marine creatures to live, hide, and hunt.

In every case, the results are the same. Soon after they are deployed to the bottom, new reef materials are blanketed by a living carpet of filter-feeding creatures such as mussels and barnacles. Crabs, snails, and shrimp soon follow, and in their turn, tens of thousands of reef fish take up residence on the reefs. The results are often spectacular, as exemplified by the following sites.

68 Rockaway Reef

Directions: Motor 4 miles out of Rockaway Inlet on a heading of 143 degrees to the Rockaway Reef.

Rockaway was the first artificial reef in New York waters. It was originally conceived by Captain Laddie Martin, who ran the well-known headboat *Rocket* out of Sheepshead Bay. Laddie's idea was to build a reef off of Rockaway. It was an idea that our friend Captain Howie Berlin took to the director of the Marine District of the State of New York. Together the three worked with the Army Corps of Engineers until the project became a reality around 1974.

The site of the reef can be found on most charts. It lies between the *Warrior* Wreck buoy and the Big Wreck inshore, and inside of the offshore marking of the *Turner* Wreck. These wrecks, it should be noted, are themselves wonderful fishing sites. Manny still has fond memories of a half-day catch of seventy-four porgies and forty-one pilotfish one summer while fishing the *Warrior* Wreck.

The Rockaway Reef was begun by dropping overboard pyramids of tires lashed together with very heavy chain, with its progenitors donating the use of their boats for transport. Later, barge loads of cleaned concrete and steel from the demolition of the railroad bridge over the Reynolds Channel were added, and the reef continues to be augmented to this day.

The payoff is pure fishing pleasure after a very easy trip from shore. The reef is loaded with ling, blackfish, sea bass, porgies, and small blues. In addition, false albacore, weakfish, and bonito roam the bait-rich area. Often fluke can be caught on the reef itself or by drifting just around its perimeter.

Tip: In early spring, blackfish have soft mouths. Consequently softer bait works better than crab. The belly of a clam will produce fish, as will 2-inch pieces of bloodworm or sandworm.

When targeting sea bass, remember that your tackle should be a bit heavier than the size of the fish might otherwise indicate. At a minimum, you will need a stiff rod and 20-pound test line, since these scrappy fish will head for the nearest bottom structure the moment they are hooked, and you need gear with enough backbone to turn them.

Figure 22 ■ John Mihale caught this "double header" of fine blackfish on the Rockaway Reef.
(Photo: Steve Heins, New York State DEC.)

69 *WAL-505*

Directions: Travel 7.8 nautical miles out of Rockaway Inlet on a course of 143 degrees to reach the site.

Heading back to the Brooklyn Navy Yard on board a destroyer escort after a tour of duty in late June of 1960, Manny was in the bow looking for *Ambrose*, a red-colored light ship that marked a channel to the east. As it happened, the *Ambrose* had been returned to the shipyard for repair, replaced by another light ship, *WAL-505*. However, this relief ship was nowhere in evidence, either. As it happened, she had sunk a day or two earlier when the 10,270-ton freighter *Green Bay* struck her in a dense fog.

To this day, *WAL-505* (also still called "Relief,") remains intact in 96 feet of water. At 129 feet long with a beam of 29 feet, the superstructure of the ship's highest point rises 30 feet and is a major feature in more than fifty chunks of artificial reef and other structure within a 5-mile radius.

The bottom fish caught in and around *WAL-505* prominently include red hake (ling), especially when the water starts to warm in May and June and then when it begins to cool in October and November. Blackfish also feed and hide in and around the wreck. Sea bass, schoolie cod, and smaller pollock also are taken here. Skimmer clam or a piece of herring or perhaps a strip of fresh mackerel are the very best baits on *WAL-505* for all of the fish that swim around her hull except for blackfish, which prefer a half of a peeled green crab. (Note: Since size limits apply to some of the fish taken here, it is important to know current regulations.)

Tip: Schools of voracious bluefish, bonito, false albacore, and mackerel pass above the wreck en route north in the spring and again while swimming south each winter. Often herring swim among them. If you happen to catch a herring, cut it into strips and get it back out into the water as quickly as possible. Freshly cut herring strips make superb bait.

Interestingly, the rich waters of this area also feature lobsters, called "bugs" by local divers. Divers working the wreck identify their presence by buoys and flags, and often take a fine dinner home from the sea.

70 *Arundo* Wreck

Directions: Sail 24.1 nautical miles out of Jones Inlet on a 204-degree course to reach the wreck of the Arundo.

Built in England and launched in 1930, the steel-hulled, 412-foot *Arundo* was torpedoed by a German submarine on May 6, 1942—not in the mid-Atlantic or off the coast of England, but right here in our own waters, just 15.5 miles from the *Ambrose* Light.

The wreck of the *Arundo* lies on the eastern edge of the Mud Hole, a vast area of deep water that is a magnet for big fish. The waters of the Mud Hole range to depths of 240 feet, which keep water temperatures constant year-round.

The *Arundo* lies in 130 feet of water and pieces of its superstructure rise as much as 40 feet. Reachable from inlets closest to New York City, the *Arundo* is a popular destination of the larger headboats out of Sheepshead Bay and Freeport—at least those licensed to sail more than 20 miles from the dock. Charter and private boats often fish these waters, too, in quest of flashy game fish and bottom-dwellers alike.

The varieties of fish taken on this wreck are mind-boggling. Bluefish, false albacore, and oceanic bonito swim these waters in great numbers; in fact, they prove a nuisance to anglers who are after bigger game. And certainly bigger game is a real possibility here. *Arundo* is visited every summer by big bluefin tuna, which appear in early July and remain for a month or so. In August, yellowfin tuna make their appearance, along with schoolie bluefins. Both remain in this rich feeding ground until the end of September, about the time the bigger tuna return to feed here for another 60 days or so.

Plenty of sharks are taken around the *Arundo*, including makos, blues, and browns. Bottom-dwellers that swell the catch include codfish, pollock, sea bass, and hake, both white and red. Big blackfish are occasionally caught here as well.

Tip: For bait, we suggest mackerel for virtually everything, especially if you can get it fresh. A chunk of Boston mackerel will entice every fish on the

Arundo to feed (other than blackfish). We have caught many codfish, sea bass, and hake on mackerel, and tuna and sharks love it, too.

Before cleaning fish, be conscious of the law. Some fish must be kept whole for possible inspection when you get back to the dock. For everything else, have a cooler at hand filled with ice.

71 Town of Hempstead Reef

Directions: Head 3.5 miles out from Jones Inlet, traveling south by southeast, to reach the reef.

The Town of Hempstead Reef is more familiarly called "Our Reef" by Skipper Al Lindroth of the *Captain Al* because he started it back in the mid-1960s. Captain Lindroth enlisted the enthusiastic cooperation of others, beginning with the then-mayor of Freeport. Captain Al Ristori, the well-known saltwater columnist for the *Newark Star-Ledger*, helped get the Garcia Company to contribute to construction of the reef. Captain Bob Kearns contributed a half-dozen surplus barges that Lindroth arranged to have towed offshore and sunk on the reef site.

Over the years, more material was added, including a load of concrete from a demolished bridge, a Navy barge, some old dry-dock waste, a commercial fishing boat, and even an old Fire Island Ferry. In addition, a bunch of cleaned-up old Army tanks were sunk here.

The result is an artificial reef that is 1 mile long and up to a third of a mile wide, a mass of structure that attracts sea life like a magnet. Not surprisingly, it attracts anglers, too. In fact, anglers in their own boats usually find "Our Reef" with ease, not because it is marked by a buoy, but because hardly a day goes by that several other boats are not fishing it.

The usual bottom fish feed on the reef, often in very significant numbers. Big ling (red hake) are caught here in April and early May and again in October and early November. Sea bass also appear in May from their wintering grounds far offshore. The bigger "blue-nosed" variety, averaging from 1.5 to 3 pounds, are taken here, too.

Many blackfish are caught on the Town of Hempstead Reef, most often on a half of a green crab or a fiddler crab.

Tip: Try using a two-hook rig that you fix up yourself rather than using a tackle-store rig. Tie a snelled hook just above your sinker knot and then tie in another snelled hook halfway down the first. This is as good a rig as you can get for blackfish.

In October, porgies flood onto the reef. While conventional wisdom is to use clams (the bait that generally will be available on headboats), the anglers who catch the most fish bring their own bait: worms. A piece of bloodworm or sandworm will produce more action.

Hooks of choice vary with the fish you are after. For example, we recommend a size #4 Virginia (blue in color) for blackfish, a size 1/0 beak style for ling and sea bass, and for small porgies, a size #4 or #6 with baitholder barbs.

"Our Reef" is a good example of something people can do in nature that is creative rather than destructive: a super fishing area created where only flat, featureless ocean bottom once existed.

72 *Resor* Wreck

Directions: Sail 49.3 nautical miles from Jones Inlet on a heading of 185 degrees to reach the R. P. Resor *Wreck.*

Built in Kearny, New Jersey, in 1936, the *R. P. Resor* is another of the many vessels sunk off our shores by German submarines during World War II. Hit by two torpedoes on February 27, 1942, the ship's cargo of oil burst into flames but remained above the horizon for some hours. The crew abandoned ship but the frigid water claimed all but two of a complement of forty-three men.

Today the *Resor* lies on bottom along the 20-fathom curve that parallels New Jersey's eastern barrier beaches. When it sank, the 445-foot ship broke into two pieces, which lie close to each other in its final berth in 122 feet of water.

Figure 23 ▪ Cod.

Big fish are caught at different levels around the *Resor*, ranging from tuna caught above to big codfish right down in the structure of the ship.

Mackerel arrive in great numbers early each year, followed by bluefish. But most anglers making this long, sometimes arduous trip are generally after the fish that dine on mackerel and blues. In early summer, when the ravenous bluefish have chased most of the mackerel north, mako sharks turn the tables and make meals of the blues. The fishing for sharks is especially good above the *Resor* at this time. (A few miles to the south, an area called the "Fingers" is another prime shark fishing venue.)

In early July, when they reach 70 degrees, the waters around the *Resor* are often alive with tuna, especially schoolie bluefins and a few yellowfins. Skipjack, bonito, bluefish, and false albacore abound and often take baits intended for the tuna.

Tip: If smaller game fish are interfering with your tuna fishing, work them into your game plan. Have a separate stand-up rod rigged with a high-quality reel and 300 yards or more of 30-pound mono. Stick a size 7/0 tuna hook through the back of your little blue and let him go fishing for you. The light line will not scare away the tuna, who may be concentrating on your bluefish trying to escape. If so, buckle your seatbelt! The next sound you hear will be screeching drag.

If you are after big codfish, rig with a single hook and a breakaway sinker. If you are not especially bothered by the need to rerig from time to time, add a second hook just above the sinker, to drop a foot below the lead. This second hook should be smaller than the top one and may produce cod as well, but is really intended for big fat-bellied white hake or ling.

A 10- to 20-pound white hake is at best an unlovely sight. However, if you have a big cooler, put the hake on ice. Properly cleaned, it will yield some fine eating.

Baits of choice for cod and hake are fresh skimmer clams. If you are on a head boat, ask the mate to show you how to bait your hooks because style does matter. Our own preference is to stick with a third of the tongue of the clam on the bottom hook and three or four such pieces on the higher codfish hook.

73 Fire Island Reef

Directions: The Fire Island Reef is located 2 nautical miles south of the Fire Island Lighthouse.

Fire Island Reef is a massive 744-acre site, which is a haven for bottom-dwelling marine creatures. It is constructed of an incredible variety of materials, including a 43-foot steel sailboat, a total of sixteen barges, two

Figure 24 ■ Old Good Humor Ice Cream trucks and tire/concrete units were dumped overboard from wooden barges in the 1960s to build artificial reefs.
(Photo: Steve Heins, New York State DEC.)

other boat hulls, seven armored vehicles, cleaned concrete of all kinds, two wooden dry docks, and at least fifteen hundred tires. Equal In size to the huge Hempstead Town Reef, the Fire Island Reef lies in 62 to 73 feet of water.

The location and accessibility of the reef, which exemplifies New York's artificial reef program, are of extreme interest to us. Other states' programs often locate artificial reefs much farther offshore, whereas most of New York's reefs are closer in and more accessible to many more fishing vessels. This makes all the sense in the world to us.

As at other reefs in New York, prime fishing time is summer, and primary species are sea bass, porgies, and blackfish. (Remember there are size and bag limits.) However, the reef produces lots of ling in early May. Ling head inshore from such deep water as the Mud Hole every spring and often stop on bottom structure like the Fire Island Reef. Ling love to hide inside tires and dart out to snatch a piece of skimmer clam as it swings by.

Tip: If you are like us, you prefer fresh fillet of ling to nearly all other fish. However, to assure a delicious dinner, you will need a cooler with ice to keep them cold from the moment they are caught. Otherwise they will soften up in your bucket.

Hook size is important when fishing a reef. As elemental as this may seem, it is important to choose a hook that fits the size fish that are being caught. If on a party boat, just check what the mates have tied onto the rental rods and match them. These guys know what to use. (Chances are, your hook will be a size #1 beak style with baitholder unless you are targeting blackfish. Then a blue blackfish hook will be the one you need.)

Regarding baits, blackfish will go after half of a green crab more than anything else on the reef. For porgies, try a clam belly if the bergals are not too numerous. Sea bass are the least finicky, but a chunk of mackerel may be at the top of their menus.

In September and October, big fluke are caught by drifting over the reef. Because of all the structure at bottom, it is especially important to tie on your sinker below the knot that holds your hook. This way, if a "doormat" drags your sinker into bottom, it can break off and the potential pool winner can still be yours.

Bluefish also make appearances over the structure, and when they do, you will be well-served to have a second rod along, just in case. A 4-ounce "Ava"-style jig can be cast out and retrieved quickly. (If you allow it to bounce along bottom, you are virtually certain to lose it.) If the boat is crowded, of course, casting artificals is not an option.

Late in the summer, exotic fish come up the Gulf Stream and make their appearances, too. Plenty of pools are taken by 2- and 3-pound trigger fish. Banded rudder fish (a.k.a. pilot fish) are caught on virtually any natural bait on the way down to the bottom. The same is true of blue runners. And what is also true of both fish is your surprise when they come out of the water. They fight so well that you will swear the fish was twice as big as it actually is. Just another small pleasure at a big and bountiful Fire Island Reef.

74 Airplane Wreck

Directions: Motor out to a point about 300 feet north of Buoy 32A, near the Middle Ground and the Stamford Reef off Stamford, Connecticut.

A Navy trainer flying out of Roosevelt Field crashed into the Long Island Sound around 1970. Soon after settling to its last resting place on the bottom of the sound, the plane's wreckage was colonized, first by tens of thousands of mussels and barnacles, then crabs, snails, and ocean-bottom crustaceans, and finally an array of hungry fish.

Our friend Captain Howie Berlin discovered the wreck in the mid-1980s and had the site pretty much to himself until it was more broadly discovered in 1992. Captain Berlin ran his headboats *Claire* and *Super Claire* out into the sound literally thousands of times, and the Airplane Wreck was a traditional destination for a six-week period each year, from October 12 to November 27. In that period, he could count on getting his anglers into loads of blackfish. The primary bait was green crabs, but fares who did not like the mess of cutting a green also caught fish on fiddler crabs.

> *Tip: When selecting a green crab for bait, take one that has some orange in its coloration. These often make the best baits.*

Some years after the plane settled to the bottom, a barge was sunk at the same site, creating even more fish-attracting structure. It is a fine site for bottom fish to this day.

75 Texas Tower

Directions: Sail 57.5 nautical miles south by southeast out of Fire Island Inlet on a heading of 163 degrees.

Say the word "Texas" and exaggerated size comes to mind. In the case of the Texas Tower, this is wholly appropriate, not only because of its large debris field but also for the size of many of the fish taken in the 186-foot depths at this hot spot.

The Texas Tower grounds are named for a USAF early-warning radar installation constructed in 1957. The name "Texas Tower" came from design similarities to Gulf Coast offshore oil-drilling rigs. In the late 1960s, the tower broke up in a severe winter storm and fell to the ocean floor, creating a debris field much larger than a football field. This giant junkyard, which included the tower's massive framework, helicopter landing pad, radar domes, and crew quarters, was later augmented with additional debris, including two barges that were scuttled at the site. All of this hard substrate material was colonized by mussels, mollusks, crabs, shrimp, small fish, and then ever-larger fish. Superb fishing soon followed at this deepwater artificial reef.

Fishing at "The Tower" is excellent year round, according to Captain Gary Fagan, whose Belmar, New Jersey–based *Big Mohawk III* is one of several headboats in the area rigged and equipped for deepwater fishing. (Coast Guard regulations prohibit most such boats from sailing beyond a 20-mile limit.)

Giant pollock and codfish are the featured fish in this location, especially in autumn and spring. As pointed out by John Raguso in *Atlantic Wrecks*, "the former New York State record pollock, a 44-pounder, was caught almost by accident by a lucky teenager while he and his dad drifted for sharks directly over The Tower's hulking submarine frame."

Captain Gary generally sails early when he fishes The Tower, since the best fishing for pollock is usually in the dark before sunup.

Tip: At such times, a 10- to 17-ounce jig works better than a baited hook. Some intrepid anglers add a very large red or green tube-teaser several feet above the jig, often leading to a hysterical tussle when two huge fish strike simultaneously. Cod begin biting at dawn. Large white hake and their smaller cousins, red hake (ling) are taken here on bait, day or night.

Anglers jigging for pollock between late August and the end of October are often stopped cold in their retrieves by bluefin tuna—generally "schoolies" in the 10- to 50-pound class. Regardless of size, tuna tear line off reels at incredible speed, and anglers at The Tower do well to keep this in mind. The Tower also attracts marlin, dolphin, yellowfin tuna, mako sharks, and even the occasional Allison tuna in their forays from the Hudson Canyon, a scant 12 miles to the southeast.

Is Texas big? No matter how you measure this "Texas," the answer is a ringing affirmative—especially in the size of fish it harbors. Just thinking about it stirs the blood: blue and white marlin, massive sharks, tuna, wahoo, outsized pollock, and cod awaiting us in those deep blue waters, far offshore where we can escape boats and crowds and, more importantly, the everyday cares of the real world.

76 Great South Bay Reefs

Directions: To reach Kismet Reef, travel 120 yards north of the South Beach west of Kismet. To find the Fisherman Reef, go 900 yards east of the Robert Moses Fixed Bridge.

Kismet Reef and the Fisherman Reef are key fishing locations in Great South Bay. These reefs are especially appreciated by anglers who want to fish "inside" when the weather is bad. To develop a more detailed understanding of the reefs, we talked to Steve Heins, who is in charge of the artificial reef program for New York State. Steve was extremely helpful, not only in providing a wealth of information about the program and the reefs themselves, but also in discussing fishing the reefs.

The larger of the two reefs in Great South Bay is Kismet Reef. Covering 10 acres, the reef is covered with a dense volume of materials, including two barges, twenty-four thousand cement blocks, all manner of concrete rubble, and some four thousand tires lashed in three-to-a-unit sets and sunk. The water here runs from 16 to 25 feet in depth; consequently, light gear can be used (providing your rod has a stiff-action tip to help you bring a blackfish up and out of its hiding place). Steve described taking sea bass, blackfish, and porgies at Kismet Reef in late summer.

The Fisherman Reef lies in slightly deeper water than Kismet, with depths varying from 25 to 40 feet. *The Fisherman* magazine was responsible for most of the effort involved in creating this additional inside hot spot; hence the name.

Covering 7 acres, the reef contains 100 concrete reef ball units, an old wooden vessel, and other debris. As *The Fisherman* magazine gets additional contributions toward this reef, it will be expanded even more. The staff and owners of the magazine are truly dedicated to providing quality fishing for all anglers, as illustrated by this effort.

These two reefs are ideal places to take children fishing. If the wind is not cranking, a trip to Kismet or The Fisherman Reef is a great way to get that young person involved in your sport. (If you are on a headboat, you might ask the mate to help you teach if you are not expert in this particular kind of fishing.)

The primary goal is bottom fishing, but weakfish may very well bite on a change of tide. The last hour of incoming or outgoing tide will often bring sea trout, which are fished very early or late in the day, since they are notoriously noise-shy.

Porgies are taken late in the summer, and a double-header of half-pound "scup" pulling in opposite directions will thrill your youngster and test his or her muscle power as well. If at all possible, fish weekdays, since boats tend to be crowded on Saturdays and Sundays.

Tip: Rig for porgies with two hooks, one tied into the snell of the other. A size #4 Eagle Claw with baitholder is the hook of choice, and your bait should be a 2-inch piece of sandworm or bloodworm. (The worms are a bit more costly than squid or clam, but they really produce results.)

When sea bass are the primary target, squid or clam will work just fine, since bass are not as finicky as porgies. They often are caught on a snelled hook that is tied above the sinker, dropping to a foot or so above the sinker knot. "Double-headers" are a common experience with sea bass. We have particularly good luck using a small piece of skimmer clam for them, baited on silver colored non-offset hooks in size #2.

Blackfish also abound on both of the Great South Bay reefs and are fun to catch. While they run smaller inside, they fight very well, and a hard-charging 2- or 3-pounder will provide fun for you as well as your youngster. To catch blackfish, many "inside" anglers prefer to fish with a whole fiddler crab on a size #4 blackfish hook. However, two hooks baited the same way will increase your odds.

77 Hylton Castle

Directions: From Fire Island Inlet, steam 8 to 9 miles out in a south by southeast path. For exact location, consult one of the fine wreck books carried by better Long Island tackle stores.

The *Hylton Castle* is thought to be a small tug, sunk many years ago and lying in sand bottom in 80 to 90 feet of water. The wreck is about 60 feet long by 25 feet or so wide and rises up about 5 feet off bottom, making it relatively easy to find on sonar if you know what to look for.

One of the true experts on the wreck is Captain Peter Jakits, who runs a 35-foot Henriques Maine Coaster called *Rainbow Runner* out of Montauk. Captain Jakits describes the *Hylton Castle* as one of his favorite small wrecks and one he fished many times when running out of Fire Island Inlet.

Because the wreck is small, it cannot accommodate a great deal of fishing action. Several boats can drift over it, but if one big boat is anchored, a second large craft cannot do more than drift its edges. Consequently it is more a weekday than a weekend spot.

Anglers enjoy super catches of large sea bass all summer here, especially when the wreck is not fished every day. Top baits are, in order of preference, skimmer clams, squid, and killies. We have caught many bass on chunks of herring or mackerel, too.

Tip: Try a top and bottom rig, with softer bait such as a piece of clam tongue on the hook above the sinker and a squid strip below. And always carry some herring or mackerel and have a jig or two in your tackle box. While natural bait catches most sea bass, Captain Jakits has taken his largest fish on the wreck with a diamond jig.

In summer and again in fall, look for bluefish above the wreck. They feed at all depths, from 10 feet below the surface to right down in the wreck, where they hunt tasty bergals.

In late fall, a run of porgies picks up the action at the castle. Porgies will take the same baits as sea bass, but their preferences run to worms. Of all baits for porgies, bloodworms, or sandworms seem to be most productive.

The *Hylton Castle* often produces meals fit for a king—tasty fillets of sea bass that rival any gourmet's delight taken off the New York coastline. And just as often, it produces a day on the salt filled with fun and excitement.

78 Tug and Tow Wreck

Directions: Sail 20 miles out of Fire Island Inlet on a 150-degree heading, traveling in a southeasterly path.

The Tug and Tow comprises two sunken vessels, the smaller a tugboat and the larger the barge it was pulling when both went down. The tug has broken down badly and does not offer as much structure as it once did. However, the barge is nearly intact and presents a flat surface that is covered with barnacles and mussels, and of course, plenty of bottom-dwelling fish that cruise past in search of an easy meal.

The water is 130 feet deep; thus the wrecks are in good, cold water where winter fish can be taken all year long. Consequently, although good fishing here begins in March, it often continues right through the warmest months of summer.

Most of the fish caught here are cod, ling, and pollock, with some blackfish and sea bass mixed in. Cod and ling are the most sought-after fish on the wreck.

Other less desirable fish also are taken here, including what some anglers incorrectly call "conger eels" (the actual name is ocean pout) and spiny dogfish, called "sand sharks," which are almost always in evidence. Captain Castaldi told us that commercial long-liners sell sand sharks at the market for upwards of $.50 a pound. The taste can be a bit strong unless they are bled quickly. If you decide to try a dish of dogfish, cut the heads and tails off as soon as they are caught.

Tip: The best rig here is a size 6/0 Sproatbend hook in black or bronze tied on a 2-foot stiff mono leader. (You can buy these pretied.) Put the snelled hook on your line with a dropper rig 4 feet above the sinker. Then add a blackfish ("Virginia") hook below the sinker on a 12-inch leader. The top hook is for cod and the smaller one below will catch ling. Bait each hook with a fresh skimmer clam sized appropriate to the hook and fish.

Ocean pout are very easy to skin and fillet and make surprisingly good table fare. If you are fishing on a head or charter boat, ask the mate to show you how.

79 *Virginia* Wreck

Directions: Sail 45 miles out of Fire Island Inlet on a course of 160 degrees.

The *Virginia* is a large ship that went down off the coast of New York many years ago. Thought to be as long as 300 feet, the ship is breaking apart but still provides a fine area to fish.

Figure 25 ▪ Roy Kneisel boated this 20-pound codfish in "tee-shirt weather" on board the *Terminator* at the *Virginia* wreck. *(Photo: Captain Tony Castaldi.)*

The *Virginia* is a true 365-day location. A good run of sea bass, including some very large fish, shows up here once the water gets too cold inshore for their comfort. Codfish are found on the *Virginia* year-round, with the action often surprisingly hot when the fishing pressure on the wreck has been light. Pollock are caught on the wreck, too. Some anglers prefer to fish for them with a heavy jig well above the structure. Others take these speedsters dead on bottom, using the same skimmer-clam-baited hook that is so attractive to cod.

The center of the ship rises to a peak of nearly 40 feet off bottom. Fishing captains refer to this kind of structure as "sticky," meaning that it causes a lot of lost tackle. Because of that (and because we do not favor those store-bought rigs that come wrapped in plastic baggies) we recommend rigging as simply as possible here.

If you use standard-prepared two-hook rigs with sinkers and barrel (or worse) snap swivels, you increase your odds of catching superstructure rather than fish. We recommend you buy a dozen or so snelled hooks to fish the *Virginia*. Tie size 5/0 to 7/0 Sproatbend black hooks onto 2 feet of 40-pound stiff mono and affix the leader to your line about 5 feet above the sinker.

If you want to fish a smaller hook below the lead, try a size 1/0 or 2/0 baitholder hook on a 12-inch leader, tied 2 inches above the sinker on a dropper loop.

Tip: With so much sinker-grabbing structure, tie your sinker on with the weakest knot possible, perhaps as bad a knot as a double overhand. If you hook a nice codfish and it drags your sinker into structure, chances are good that the lead will snap off at the weak knot and you can still bring your fish to gaff.

In addition to codfish, pollock, and sea bass, anglers fishing the *Virginia* often take ling (red hake) and their larger cousins, white hake. Ocean pout are caught here in numbers, too.

80 *Coimbra* Wreck

Directions: Sail 27 nautical miles from Shinnecock Inlet, on a course of 183 degrees.

The first two sinkings of ships by German submarines off our shores occurred in January 1942. The first was the *Norness*, sent to the bottom on January 14 by U-boat 123. The following day, the same sub attacked the *Coimbra* about 20 miles from Shinnecock.

The *Coimbra* was a 423-foot tanker filled with oil when hit by a single torpedo, which set the ship ablaze. Breaking up into three pieces, the burning ship soon sank, taking thirty-four of her forty-four crew members to their deaths.

Today the wreck rests in 173 feet of water, with pieces standing as high as 40 feet off bottom. Divers report the steel hull has deteriorated little over the years. The ship's middle section rests partially buried in bottom, but the fore and aft sections are completely exposed and covered with filter-feeding organisms and crustaceans, thus providing a great area to entice bottom-dwelling fish to feed.

Our friend Captain John Raguso, writer of the authoritative book *Atlantic Wrecks, Book One*, catches monster codfish and pollock at the *Coimbra* virtually year-round. Weather permitting, you can generally count on good catches here even in the hot summer months if the wreck hasn't been overfished. Being in such deep water means the *Coimbra* is a comfortable feeding ground for these "winter fish," regardless of surface temperatures.

The 30-fathom curve lies near the wreck, and Capt. Raguso has caught mako shark here while drifting bait. Tuna and white marlin also are nailed at or near the *Coimbra* in summer.

Tip: When fishing a wreck like the Coimbra, *pollock are usually caught well off bottom, or on the edges of the structure, rather than down into the interior. Try using a large diamond jig with the treble hook removed and replaced by a single size 6/0 silver non-offset hook (to avoid losing expensive tackle) using a strong split ring. A big pollock can be hooked on one barbed hook as well as three barbs of the treble.*

The cod are found right in the wreck itself, so you will need plenty of extra terminal tackle. Use just one hook and tie it on a 3-foot leader, 4 or 5 feet above the sinker. Make sure your sinker knot is weak so it can break off if a big "winter king" swims into structure and hangs your lead.

Close-in Waters

Great saltwater fishing is often surprisingly close at hand. New York's bays, inlets, and jetties are often remarkably rich in marine life. Fished from boat or shore, pier or jetty, the following are among the best of a wide spectrum of wonderful choices.

81 Raritan Bay

Directions: Motor 9 miles west by southwest from Sheepshead Bay to reach Raritan Bay.

Raritan Bay is an extremely large body of water, mingling waters from the outfall of the Raritan River in New Jersey to the west with waters of Lower New York Bay to the north and east. It is a fine resource for anglers bent on striped bass, bluefish, flounder, and fluke. April through June are among the best months to fish the bay, especially for flounder and schoolie bass. Particular hot spots are Princess Bay, Seguine Point (near Lemon Creek), and the Conference House.

Surf casters begin to score on striped bass in late March, and tons of schoolie stripers appear in the bay in April. Anglers score well on the short bass with sandworms fished just off bottom.

Figure 26 ■ If you caught a 20-plus-pound striper in New York City waters as this young man did, you would smile, too. *(Photo: Bernie's Fishing Tackle, Brooklyn.)*

Captain Kevin Bradshaw of the *Dorothy B VIII* out of Sheepshead Bay fishes Raritan Bay every spring for flounder. He fishes the edge of buoy 26, the red side of Raritan Reach and the green side, south of Raritan Reach. Closer to shore, he favors Keyport and Keansburg.

Captain Bradshaw likes to use chum pots filled with broken up mussels and steamer clams. Light chumming with corn kernels also improves the catch. We always thought it was simply the bright color of the corn that attracted flatties, but Kevin told us that the corn simulates steamer clam necks. It turns out that flounders are fond of nipping the necks off young steamers when they stick them out of their shells on bottom.

Fluke action improves in the bay with the warmer weather in May, and fluke are taken in the same waters you have been fishing for flounder. The *Dorothy B VIII* often sails up into Princess Bay and along both sides of the Raritan Reach in May and June. Obviously the edges of any channel buoy mark a potential holding area for fluke. If you are skilled at anchoring and can position dead on the edge of the channel marker, your boat should dust back and forth over a distinct depth change—and more than incidentally, over just the spot where some of the biggest summer flounder wait to ambush passing food.

Also in spring, bluefish from 1 to 10 pounds and nice weakfish come into the bay. The best times to catch them are early each morning and at tide changes. The "weakies" are caught on sandworms.

The bay continues to produce right through summer. Incoming tides produce super fishing for fluke, blues, and weakfish in the hottest part of the summer. As the water begins to cool in the fall, blackfishing heats up, with particularly good catches recorded at the Staten Island Monastery. Huge bluefish go for peanut-sized live bunker at this time, and stripers are taken on live eels, bunker, herring, or plugs.

Shore fishing spots abound from piers, jetties, and bridges, and surf casters score well from the beaches that fringe this wonderfully productive bay.

82 Gateway National Recreation Area

Directions: Take U.S. Route 278 across Staten Island to Father Capodanno Boulevard and the Gateway National Recreation Area (NRA). Or take Route 27 in Brooklyn to Shore Parkway to the Jamaica Bay–Breezy Point Unit. The Gateway NRA also is accessible by subway or bus.

Gateway NRA comprises three separate units in the New York City harbor area: the Jamaica Bay–Breezy Point Unit, the Staten Island Unit, and the Sandy Hook Unit. The best fishing of all the units may be at Sandy Hook, which is a prime spot for striped bass.

Sandy Hook provides many classic in-shore fishing spots. Our personal favorite is Sandy Hook Bay, where action starts to heat up in early spring and never cools down until the coldest winter months.

Early in April, anglers baiting with skimmer clam, herring, or mackerel strips start to catch ling in significant numbers just inside the bay at buoy 1. Later in April, hungry flounder in large numbers leave the rivers and wintering holes in the bay and start to feed voraciously on offerings of sandworms, bloodworms, and clams. Special flounder hot spots include the Pump House near the Leonardo Flats, Spermacetti Cove, and Plum Island at the mouth of the Shrewsbury River. Another dynamite spot for huge April flounder is the Oil Dock between the Navy Pier and the Atlantic Highlands Marina.

Stripers begin to arrive in Sandy Hook Bay about the time the magnolia trees start to blossom, usually in mid-April. Before the month is out, some fluke also appear in the bay.

Memorial Day is a reliable time to begin drifting the bay for fluke, and in June fluke in large numbers start to get serious, especially across the bay at Officers' Row and around the Pound Nets. Trolling big spoons and umbrella rigs during the day (and plugs at night), especially out near the end of the bay at the rips, are good ways to catch stripers.

Super fluke catches began with the tee-shirt weather of July, especially along the drop-offs at Pound Nets and Officers' Row. Standard baits are recommended, but anglers also have luck here with a white 3-inch Mister

Twister and with a #1 nonweighted white bucktail and small killy, held 18 inches above the sinker on a 6-inch leader.

More great fluke fishing is found at the drop-offs right in front of the Navy Pier on the red 2 and 4 channel and Terminal Channel buoys 2 and 4 red or 1 and 3 green.

Summertime signals showings of stripers and weakfish in Sandy Hook Bay, the former from Horseshoe Cove to the Bug Light; the latter around the Navy Pier and near Leonardo Marina. Effective offerings for weakfish generally involve sandworms, either plain or "à la mode" (that is, combined with a strawberry jelly worm combo or other concoction). The bay also is loaded with snapper blues all summer.

In October the striped bass is king of the bay, with trophy stripers taken while trolling bunker spoons near the Pound Nets, the Bug Light, and the channel leading into the Shrewsbury. Headboats score well on line-sides in the evening on incoming tides, especially at Spermacetti and Horse-shoe Coves, with live eels or extremely large sandworms. Fat, hungry flounder also are active from October through December, rounding out the calendar year in this great fishing area.

83 Hoffman Island Sand Bar

Directions: From the mouth of the East River, steam 190 degrees to buoy 10, then turn southeast to Ambrose Channel.

There are those fishing spots we love not because of huge, tackle-busting tuna or flashy striped bass or even great grab bags of outsized bottom fish. There are those spots that simply provide lots of action on a regular basis; action that puts fish in the cooler and fills the memory with wonderful days on the water.

Such a spot is the Hoffman Island Sand Bar, located off Hoffman Island on the Staten Island side of The Narrows, south by east of Fort Wadsworth. This sand bar was discovered by Captain Howie Berlin many years ago and became his favorite spot in the area, especially in the early part of the season.

What he found on the sand bar were fluke, literally tons of 2-pounders, in very shallow water, especially in early July. The hungry fluke come into the shallows early because this is where the water is warmest.

The sand bar is essentially a shelf at the north end of the island and it is easy to locate with any reliable depth finder. Area charts also pinpoint it. Captain Berlin also likes the green side of Ambrose Channel, at and west of buoy 19.

Because it is close in and because of the wonderful action, the Hoffman Island Sand Bar is a perfect location to take the kids. Catching one fluke after another from a comfortable boat on a calm sea is a sure-fire first step toward building fishing partnerships for life!

84 Norton's Point

Directions: Take Route 278 to Coney Island. Norton's point lies just off of Coney Island.

Norton's Point affords the angler a fine, close-up view of many New York landmarks as well as wonderful fishing for fluke and weakfish. When facing landward from offshore of the point, Coney Island is dead ahead, the Marine Parkway Bridge on the right, and New York Harbor, Staten Island, and the Verrazano-Narrows Bridge are on your left.

The point is part of the Gateway National Recreation Area and is a favorite destination for New York anglers interested in wonderful fishing close to shore. The starting point for many is Coney Island Flats, an especially fine spot for fluke in summer. Gravesend Bay, which marks the easterly edge of New York Harbor, offers superb fluke fishing in summer to the left of Norton's Point.

As the waters of the bay start to warm in spring and as they cool in the fall, flounder also are extremely active in the bay. Chumming with crushed mussels will produce fine catches of tasty flatties in both seasons.

However, the primary target at Norton's Point is game fish, with striped bass, bluefish, and weakfish being the targets of choice.

Striped bass like quiet water, too, but are not as picky as weaks. Bluefish, the notorious "eating machines" of the area, will hit at any time of the day or night. The old reliable "diamond jig" will work well for all three species when jigged up and down, but we prefer the newer version, commonly called an "Ava style." The sides of the Ava are not as severely sharp as the "diamond," changing the action of the flutter as it rises and falls. "Umbrella rigs" often are trolled to good effect, especially for blues and stripers.

Bucktails jigged near bottom also work well. Go with the lightest weight you can based on tide movement and weight of your tackle. In moderate tide, a half-ounce bullet-shaped white jig, tipped with a small plastic chartreuse tail, will often produce fluke as well as weakfish.

If your preference runs to baitfishing, concentrate on chunk bait first. If you have fresh herring, mackerel, or bunker, sit on anchor and chum with ground bunker as well as bits of fish. Normally bluefish will be attracted, but you will also take stripers and weakfish on chunk baits. And if something like a fast freight train takes your bait, charging around and under the boat at incredible speed without ever jumping, you have probably hooked an oceanic bonito or a false albacore. (The freshwater equivalent to the fight of an oceanic bonito is the hybrid bass, our personal favorite sweetwater fish and one that would send the pulses racing of even the most confirmed saltwater angler.)

Striped bass anglers score well in the area of Gravesend Bay and Norton's Point by drifting live eels, hooked through both lips, from underneath the bottom through the top.

85 Marine Parkway Bridge Area

Directions: Take the Shore Parkway in Brooklyn to the Floyd Bennett Field exit; follow roadway past Floyd Bennett to the Marine Parkway Bridge.

The Marine Parkway Bridge connects Brooklyn to Far Rockaway and marks the entrance into Jamaica Bay. The bridge is the gateway to many famous shoreline fishing spots, from Atlantic Beach and Long Beach to the famous Breezy Point Jetty (discussed separately.) To get the most updated information on this fabled spot, we talked to Frank Rupert of Bernie's Bait and Tackle in Sheepshead Bay.

Anglers fish in and around the bridge stantions for blackfish, scoring best on green crab. Half a bucket of keeper blacks is by no means uncommon here. Sea bass and porgies also are caught around the bridge in serious numbers, most productively on skimmer clams. In addition, from mid-spring to late fall, you may well catch flounder and fluke here. Frank told us that bluefish swing through as well, as do striped bass, which are generally caught in dark-water times.

Many years ago, we used to tie up to the old pilings to fish; however, it is illegal to attach a line to the pier stanchions today. To fish here, you must put your engine on idle. (On a change of tide, this is not difficult to do.) What you also can do is drop an anchor close to a stanchion, combined with a flotation device that can bring you into the bridge, provided there is no boat traffic. However, this should not be tried on crowded weekends, when other boats can make anchoring here quite dangerous.

One interesting phenomenon of fishing here is seeing the "flying fish" that rise up out of the water above your boat and just keep going. Actually, the fish have a bit of help from anglers high overhead who are fishing from the bridge. (This is allowed so long as the anglers are not in the lift section of the bridge.)

The bridge also marks other fine fishing grounds. As you approach it from Sheepshead Bay, for example, you will first reach Dead Horse Bay, which is part of Barren Island. This is a wonderful area to fish for flounder and fluke.

Once you pass under the bridge from Sheepshead Bay, you will see Floyd Bennett Field on the left (Brooklyn) side of Jamaica Bay. The old sea plane ramps will appear—all three wonderful places to fish. From evening until early morning, striped bass can be caught at the ramps with a Cotton Cordell Red Fin or a Gag's Mombo Minnow, Frank Rupert's favorite lure. Spring and fall, a bunker chunk or live eel also will produce bass, especially on an outgoing tide.

Between the first ramp and the bridge is a great flounder spot. Very close to shore, the depth drops off to 10 feet with a sandy bottom. Another steep drop occurs and mussel beds are found at bottom in 20 to 23 feet of water. Sandworms and sedge mussels are fine baits early in the spring. Later, harder baits such as skimmer clam strips or bloodworms catch more flatties.

Fluke also are caught on both sides of the bridge while drifting. But the ramps are super spots.

86 Breezy Point Jetty

Directions: From Brooklyn, cross over the Marine Parkway Bridge and hang a right into Breezy Point. The pass you will need to enter can be obtained at Fort Tilden. It will allow you to enter any of the Gateway National Park locations.

One of the truly famous fishing areas in New York State is this old jetty, built to separate the waters of Brooklyn and Far Rockaway. For many years boats would head out of Sheepshead Bay, and if the wind was blowing from the east, a turn in that direction around the tip of the jetty would bring anglers into much calmer waters close to shore. (As noted in our Artificial Reef section, Rockaway Reef—one of the best of the artificial reefs—is found nearby.)

Fishing the jetty itself is not for the weak of heart or the unsteady of foot. The rocks are slippery and dangerous, especially for the lone angler. We suggest you fish here with a companion, in case you fall and need help. Also, wear reliable walking shoes or even better, a set of "creepers."

A four-wheel-drive parking permit can get you close to the start of the jetty. Otherwise, a walk of about a mile is required from the spot where you park your car.

That said, the precautions and the effort are more than justified by the fishing. As Frank Rupert notes, the jetty is a natural hangout for fish large and small; certainly blackfish very close to the end of the jetty, in and around the rocks, and fluke, blues, and striped bass from one end of the jetty to the other. Porgies and sea bass are commonly caught here, too, especially on sandworms, bloodworms, or skimmer clams. To add to this wonderful variety, a run of mackerel will occur at the end of the jetty. Frank recommends a 1.5-ounce Smilin' Bill bucktail with a red-painted mouth for blues and striped bass. He adds a white pork rind strip or a dark jelly worm for added attraction.

Anglers casting lures do far better than bait anglers, but if the jetty is not crowded, you might work a second line baited with a chunk of bunker or a hunk of mackerel or butterfish. However, this is impractical when others are casting close by, especially when tides are strong, since the lures will inevitably tangle the baited lines.

Early in the spring, blackfish are taken on clam baits very close to the rocks at the end of the jetty, in some 20 feet of water. However, as the water warms a bit and the "tog" get into harder baits, a fiddler crab or half a green clam is far better.

In summer, anglers casting and retrieving from the jetty catch fluke. The preferred terminal tackle is a 3-foot leader on a fish-finder rig, baited with a killy and a strip of squid.

If you plan to fish in the vicinity of the jetty by boat, you can reach it by sailing only 2 miles or so out of Sheepshead Bay, but boaters should be aware that jetty-bound anglers are less than happy to see boats within casting range. In fact, it is not terribly fair for a boater to fish close to the jetty, since a boat can fish in any of hundreds of places, but the "jetty jockeys" are locked into their hunks of rock.

Two excellent nearby fishing spots that should be of interest in this context are Louie's Pier and Kennedy Restaurant Pier—both on private land and unavailable to other anglers, but open to anglers with boats.

87 City Island

Directions: City Island marks the entrance to Long Island Sound and is found between Fort Totten and Fort Sklar.

City Island is part of the Borough of the Bronx, although it may appear to be a small city unto itself. It features many fine restaurants and tourist attractions and, of greatest interest to us, a fleet of fishing boats.

In this review, City Island is actually the point of departure for a number of spots in Long Island Sound. We should observe that these spots are known best to the captains and guides who fish them on a regular basis. Consequently we turned to one such authority, Captain Howie Berlin of the *Super Clair*, who ran a headboat out of City Island for many years, to discuss precise locations, fish, and methods.

To reach the first of these locations, find buoy 48 only 15 or 20 minutes from the dock, and then head northeast to Execution Light, keeping the Stepping Stone Lighthouse on your starboard side. Here off "Execution Rocks" you will find the best blackfishing in the area. In the spring, when the mouths of these "togs" are tender, bait with sandworms or clams (including steamer clams), but in the fall, when they toughen up, switch to green or fiddler crabs. The last half of each tide is the best time to fish this area. Be alert for other opportunities here as well. Bluefish and monster sea bass are commonly found here too.

Incidentally, local lore has it that "Execution Rocks" got its name from a practice by the original inhabitants of the area of tying prisoners to the rocks at low tide, then allowing the incoming tide to "execute" them. Situated at buoy 44A, this area slopes steeply, from as shallow as 3 feet to as deep as 100 feet of water.

From Execution Rocks, go southwest for only a few minutes to "Gangway Rocks." Winter flounder are caught here in 5 to 15 feet of water and blackfish are taken in the rocks located between buoy 27 and a beacon buoy that marks a rock pile. (Needless to say, this is more a small-boat than a headboat site.)

Another hot spot is found to the northeast at buoy 23 off of Prospect Point. Here a drift from the southwest to the northeast over rocky, steeply sloping bottom often provides wonderful fluke action.

For porgies, flounder, and blackfish, head south of Prospect Point to Glen Cove in Hempstead Harbor. You will find the "flatties" on sand bottom, with the porgies and "tog" in the rocks. This section is only 50 to 300 feet from shore and is more easily fished from a small boat. The red buoys 2, 4, 6, and 8 mark the right side of the channel into the harbor and also mark the best fishing. Fish to the right of the buoys, on the shallow side.

Further along the sound to the east is a reef 200 feet south of buoy 21 at Matinecock Point. This reef, which slants toward the nearby beach, is an excellent location for flounder, sea bass, blackfish, porgies, blues and, often, weakfish. Fluke also are taken each summer offshore of this buoy.

The final location is a tugboat wreck, a few minutes west by south of buoy 21. This is a blackfish hot spot and is not to be missed.

88 Western Long Island Sound

Directions: Take Route 95 in the Bronx to Pelham Road and boat launches at the midpoint of western Long Island Sound, or in Nassau County, take Route 495 to Port Washington and the Manhasset Bay entry to the sound.

The western end of Long Island Sound gets a great deal less publicity in connection with fishing than the more popular eastern end. However, this part of the sound is filled with bays, quiet waters, and often superb fishing action.

An authority on this fishery is Rich Tenreiro of R&G Bait and Tackle in Port Washington. Rich spends more time fishing than do both of the authors combined (that's a lot of fishing), much of it in the western sound in an area running from the Throg's Neck Bridge east to Matinecock Point. This is a distance of about 10 miles, with a width graduating from a few hundred yards at Throg's Neck to as much as 3 miles from Matinecock Point across to Larchmont.

Figure 27 ■ Western Long Island Sound holds doormat-sized fluke like these two beauties caught by Gary Benkert.
(Photo: R&G Bait and Tackle.)

Figure 28 ■ Rich Tenreiro of R&G Bait and Tackle released this fine striper he caught in Manhasset Bay in the sound.
(Photo: R&G Bait and Tackle.)

In this part of the sound, the boundaries of four counties intersect: Bronx, Westchester, Nassau, and Queens. The number-one fish here is striped bass. Each spring, there is a truly wonderful striper run in the sound. Sometimes these runs are extremely early. For example, in 1998, the first good catch of schoolie stripers occurred in this area on February 10—considerably earlier than most New Yorkers found linesides anywhere else in the state.

What starts the bass action early each year is the appearance of spearing spawning in the shallow waters of the bays. A run of bunker (menhadden), the striper food of choice, generally starts about the first of April. Bigger bass prefer larger bait, and keeper-sized linesides get more aggressive in mid-April as they chase the bunker.

The bunker stay in the sound until late fall, feeding on mossbunker. Many of these bass remain to winter over in the sound, joined by others that come in from the Hudson.

The bays that are most popular are the shallowest, including Little Neck Bay, Manhasset Bay, and Hempstead Harbor. The terminal tackle here in early spring is light, perhaps a bucktail tipped with a 4-inch rubber shad in chartreuse, light green, or white, to match the spearing. Once the bunker run is on, you will need to go heavier.

With warmer weather in May and June, the bigger fish get more active, and the place to look for them is in channels ranging from 30 to 55 feet deep in the middle of the sound.

Two favorite spots for big bass are the mouth of Manhasset Bay and the area in front of Fort Totten.

In the fall, blackfish feed very actively at Execution Lighthouse between Manhasset Bay and Hempstead Harbor, about mid-sound. Crab bait is best in fall, but you need to change tactics when the fish run again in the spring around Hewlett and Barker Points. In spring, a piece of sandworm works wonders for the soft-mouthed blacks.

Super winter flounder action takes place in the back bays each April, and tons of fluke flood into the sound each summer—often ignored because of the great angler concentration on striped bass. Matinecock Point is a great spot for fluke. Go with a 1-ounce bucktail tipped with a squid strip and spearing combination. A strip of bunker or herring will produce big fluke.

Beach anglers find plenty of action on stripers. And the western sound also is a site of thrashing, crashing bluefish blasting through schools of bunker.

With big stripers, blues, blacks, and fluke abounding, it's easy to understand why Rich Tenreiro and others around the Western Long Island Sound can't stay off the water.

89 East Reynolds Channel

Directions: Head to the easternmost part of Reynolds Channel, out of Point Lookout just before Jones Inlet.

Reynolds Channel is one of those ideal saltwater destinations to take the kids. It is relatively sheltered from weather. It is a short, comfortable ride from the dock. And it reliably produces fish—often large fish in very considerable numbers—that fight well and provide wonderful table fare. The great majority of these fish, it should be added, are decidedly flat in shape.

One of the real experts on Reynolds Channel is Captain Dennis Kanyuk, owner of an all-aluminum half-day headboat, *Super Hawk*, which in spring through fall sails from the Town of Hempstead East Marina in Point Lookout. Captain Kanyuk told us that he has really excellent fishing in the channel every year only a few minutes out of his marina. Running from the old Doxee Sea Clam Factory west to the Lido Beach Golf Bridge, he finds tons of summer flounder for his fares. The *Super Hawk* catches some of its biggest fluke just east of the Point Lookout Bridge.

The fluke fishing really heats up early in May in the quiet waters of the channel, and at least half of these early-season fluke are large enough to take home. (As the season progresses, the percentage drops to about one quarter that can be put in the pail.)

Figure 29 ■ Big fluke are netted in Reynolds Channel on the *Super Hawk* out of Point Lookout.
(Photo: Captain Dennis Kanyuk.)

Figure 30 ■ Chilly weather produces big fluke just outside the inlet on the half-day headboat *Super Hawk.*
(Photo: Captain Dennis Kanyuk.)

A strip of fluke belly bottom with spearing is the standard bait, and usually the most successful. Certainly a cleanly cut tapered strip of squid can be used instead of the fluke strip, but we agree with Captain Kanyuk that the belly is tops. (In New Jersey it is illegal to do this, but New York allows anglers to fillet out the white meat from a keeper fluke, providing they bring the rest of the fish back whole to prove a legal-sized fish was cut.)

Around Memorial Day, 1- to 3-pound "cocktail" blues invade the inlet, wreaking havoc on smaller fish and flounder tackle alike. If you are blasted all at once and the rod is nearly torn from your hands, then the line goes slack and you reel up a line with no hook or sinker, it was not a "doormat" fluke that struck your bait, it was a bluefish!

There is nothing fancy about this fishing or our description of it. No fancy "glow in the dark" rigs, no rocket science electronics, no strategies or tricks. Whether from your boat or a headboat, it is simply a fine place to catch fish.

90 Great South Bay

Directions: Take Route 27A in Babylon to the Robert Moses Causeway. This will put you directly over the bay.

If you are looking for the best bet in New York for an "inside" saltwater spot to fish, chances are Great South Bay is the place. This is the largest body of enclosed water on Long Island, reaching more than 30 miles east to west, and ranging out as much as 5 miles. Great South also is the deepest of the bays, with depths of 10 to 15 feet widely distributed (although the north sides of East and West Fire Island are much shallower). Deeper yet are the waters where two artificial reefs (discussed separately) are located.

Both reefs provide fine fishing for porgies, sea bass, blackfish, and other bottom-dwellers. If action is your goal and size is somewhat less important, hop on a headboat, a charter boat, or your own boat and make for the reefs.

Fluke and flounder make up the majority of the catch in other waters of the bay, but marauding blues also wander in on slack tides. At night, striped bass and weakfish are caught here as well. The flounder are caught throughout the bay in cold water, and as the winter flounder start to make their run out to sea in May and June, their summer cousins, fluke, are on their way in.

In his excellent book *Good Fishing Close to New York City*, author Jim Capossela describes a short but intense run of flounder each fall in October and November in Great South Bay. These fish, fresh in from the ocean, are wonderfully clean and free of the muddy, fishy taste often present in flatties just popping out of hibernation each March. Consequently fall is the time to catch the best-tasting flounders.

The flounder season usually begins in the bay at Heckscher Flats, a flat-bottomed section located southeast of Nicolls Point, close to the Connetquot River outfall. Use a spinning or bait-casting outfit, but remember to have at least one rod with a stiff tip to deal with tides that may necessitate heavy (4- to 6-ounce) sinkers.

From the bay you can reach the ocean via the Fire Island Inlet to the west, or through Narrow Bay to Moriches Inlet to the east. The large headboat and charter-boat fleet in Captree has a short run to many of the hot areas of the bay, and their skippers know just where to take anglers for a half-day of great fun in relatively quiet waters with fluke and flounder.

The Babylon Cut is on the western edge of the bay, and good catches of striped bass and weakfish are scored here. Nearby, at the Robert Moses Causeway, anglers take good catches of flounders and blues. We are reliably informed that sea trout (weakfish) catches are often good to the east as well, at the east-west channel near East and West Fire Islands.

Still more flounder and sea trout are caught at Snake Hill Channel not far from Fire Island Inlet. And as the flounders are leaving, this is one of the prime places to try for fluke on the way in. And speaking of trout, it stands to reason that some sea-run browns and rainbows might be taken at the junction of the Connetquot River and the bay. Elsewhere we describe the superb fishing in the Connetquot for rainbows as well as browns and brookies, and of course, the Connetquot deposits fish as well as water into the bay. We know that when trout have the opportunity to go to sea they almost inevitably do so, but that they are programmed to return. Our best bet is that more than a few big returning trout are caught on slack water right near the mouth of the river—by anglers who aren't spreading the word.

Perhaps one of these days someone will find a secret holding area for the arriving sea-run trout. It you find it, will you tell? Right!

91 Jones Inlet, Outside

Directions: Take the Meadowbrook State Parkway to Jones Beach State Park; turn right on Loop Parkway to the boat launch at Point Lookout.

Just outside of Jones Inlet, there is an area stretching from the first drop-off to about 2 miles out that regularly features fine fluke fishing. Skipper Dennis Kanyuk begins the season in Reynolds Channel, but once the action heats up in the ocean, he runs his party boat outside the inlet where the fluke run large and fight hard.

Near the end of June, this area is especially productive in 30 to 40 feet of water. The temperature of the water is the controlling factor. When the water is colder, you may need to fish farther out (in warmer water). Conversely, the fishing improves closer in as spring turns to summer.

Some of the best fluke action is found at what used to be called "The Pink Hotel" off Lido Beach to the west of the inlet. (The hotel later received a coat of gray paint, but is still occasionally called by its original name.)

A pipeline running for about a mile off the beach east of the Jones Beach Tower carries clean effluent from a sewer treatment plant. Some sea bass are caught right around the pipe and here, too, the fluke fishing can be excellent.

The fishing outside of Jones Inlet is usually a half-day—a fine fishing trip for the little ones if the winds are light. Don't forget (1) to bring sandwiches and plenty of snacks, (2) to have a cooler at hand to keep your catch perfectly iced, and (3) to tip the mate for filleting your catch. All in all, you might be heading for a perfect day outside Jones Inlet.

92 Fire Island Surf

Directions: Access by car is limited to the eastern and western ends of the Fire Island National Seashore via William Floyd Parkway and Robert Moses Causeway, respectively. Ferries operate from Bayshore, Sayville, and Patchogue from May to November.

Fire Island is a barrier island off the south shore of Long Island. Looking seaward, the island features 32 miles of Atlantic Ocean beach. Looking landward, the island traces the outlines of Great South Bay and Moriches Bay, two shallow, exceptionally fish-rich bays. Relatively undeveloped, given its proximity to New York City, the island features miles of beach, marshes, and dunes, which are a balm to the eyes and spirits of urban outdoors enthusiasts.

Fire Island is not for the surf angler unwilling to walk a bit, since automobile access to the island's 32 miles of Atlantic Ocean beach is limited. However, the walk is more than worth the effort, and the angler often will return to the car encumbered by the weight of fish.

Fish are drawn into shallower water each summer and fall by the availability of abundant food. Late each summer, the small mullet that abound in the surf trigger weeks of superb fishing. Stripers and great schools of migrating bluefish storm into the surf in pursuit of the baitfish and the feeding orgy is on.

Live mullet can be bought in some bait shops but many anglers catch their own. However, many of the real "pros" along Fire Island National Seashore stick to artificials. For stripers and weakfish, especially in the fall, the lure of choice is a 5- to 7-inch yellow-and-black Bomber. The lure is made even more deadly by tying a small nonweighted bucktail teaser 2 feet above the plug, on a short, stiff leader. (The plug at the end of your line appears to be a small fish chasing an even smaller fish—the teaser.)

Blues hit metal here, preferring smaller to larger lures. Instead of the typical Ava-style 47 lures commonly used on party boats for blues, go with the smaller 007 size (with no tube or twister on the end) or a small Hopkins spoon in the surf.

Surf fishing on Fire Island hits its peak from mid-September to mid-November when the fish concentrate closer to shore. Marauding bluefish are most common at this time, providing unending pleasures for anglers who've come to fight. As writer John Hersey said in his book on the species, "Blues strike like blacksmiths' hammers." Striped bass and weakfish also abound in the surf in fall, especially around Moriches Inlet and Fire Island Inlet, both of which serve as highways in and out of the bay system for the fish.

Not surprisingly, fishing pressure increases substantially in the fall, but with 32 miles of beach, there is ample room.

Regarding access, four-wheel-drive vehicles are required to reach much of the beach area. Driving four-wheelers on the island is permitted only from September 15 to December 15 and special permits are required. These permits may be obtained at the Smith Point Ranger Station at the end of the William Floyd Parkway on the island's east end.

93 Moriches Bay

Directions: Follow the Montauk Highway east to the town of East Moriches and the Hart Cove entry into Moriches Bay.

Moriches Bay starts at Smith Point Bridge at the western end of Great South Bay and runs to the east to Westhampton Bridge. A small canal between Great South Beach and Cupsogue Beach connects it to Shinnecock Bay.

Much of the bay is sheltered from wind or otherwise protected. Consequently it is a small-boater's haven, with a myriad of fishing opportunities available to owners of small craft. Loads of flounders are caught in the bay, plus scrappy kingfish, big fluke, and even bigger striped bass. Captain Peter Jakits, who fishes out of Montauk, tells us that the wonderful fishery we enjoyed here in the 1950s continues strong to this day.

Figure 31 ▪ The 195-foot steel barge *Jean Elizabeth* was scuttled to become part of Moriches Reef in 1995.
(Photo: Steve Heins, New York State DEC.)

One of the spots we remember, the Quogue Canal near the eastern tip of the bay, is Pete's favorite early-season flounder area. Nice catches of flatfish are registered in mid-February during mild winters. However, it is important to check the regulations for the date when flounder fishing can begin. These were established in the late 1990s to protect the flounder-spawning season, which takes place in late winter.

The shallow, muddy waters of the Quogue Canal are great in as little as 2 to 3 feet of water near the bulkheads. A two-week flounder run can nearly always be counted on beginning the third week in March, before the flounder run into the bay proper, where they stay and feed until June.

The north side of Moriches Inlet has a protruding island that is another prime location for flounder and kingfish, especially on the top of the tide and the beginning of the outgoing.

The deeper channels of the south side of Moriches Bay have a fine striper fishery, and Pete Jakits tells us a live kingfish is the best of all baits for big linesides. He suggests you live-line it when the tide is light, or hold it down at bottom on a fish finder or egg-sinker rig. Go with a Mustad baitholder hook size 6/0 in brown color or a tuna-style hook.

Try for bass at the east or west cut in the inlet from late May clear into late November, working most seriously on the outgoing tide.

From early May into early October, you will find fluke in the same water where you caught flounder a few months earlier. For fluke you'll want to drift instead of anchoring, running your drift from shallow to deep. Once you reach the deepest parts of the channels—not much more than 7 feet—the fluke action will slow considerably.

When the water is slow, try a quarter-ounce white or red-and-white bucktail. Bounce the bucktail on bottom at high tide. Tip your artificial with a strip of squid and be sure to have a net handy. Big fluke are easy to lose on a regular hook, especially at the moment they are about to be lifted out of the water. The weight of your bucktail can lead to an even worse loss-to-catch ratio without a net.

But remember not to use a net for obviously undersized flatties, since you can damage them by removing protective slime in the net.

94 Shinnecock Bay

Directions: Take the Montauk Highway to boat-launch sites around Shinnecock Bay, including East Quogue, Tiana, and Newtown.

Shinnecock Bay is between 8 and 9 miles long and is the first major bay located west of the Hamptons. The bay is virtually split in half by Ponquogue Point at its barrier beach. Smaller than Great South Bay, Shinnecock Bay is a better bet for small private ("tin") boats. The fishing in the bay is often outstanding. The best action is usually found toward the eastern end. Striped bass are often caught around the channel bringing water from the inlet to the canal. At the eastern edge of the bay, near the Shinnecock Indian

Figure 32 ■ Karen Brown, Sue Doulose, and Judith Schoerlin hold 8-, 5.5-, and 6.5-pound fluke they nailed on squid and spearing half a mile outside of Shinnecock Inlet.
(Photo: Judith Schoerlin.)

Reservation, is a spot called "The Basket," which features deeper water. Flounder are active here earlier than in most other parts of the bay.

The waters near the reservation average 10 feet in depth, and a rowboat fleet fishes north of the Coast Guard Station as well as in the nearby channel. Fluke are taken in this area in late spring and summer. We strongly suggest having more than one rod with you in late spring, each rigged differently. At a minimum, one rod should be set up for flounder and baited with bloodworm or sandworm pieces and a second should be rigged for fluke. More than incidentally, while most anglers prefer drifting for summer flounder, we like to fish on anchor, a system that works for us and will work for you when done just right.

At low tide, large portions of Shinnecock Bay are less than 6 feet deep. Do not disregard these conditions. If the sun is out and it is early in the season, some of your best fluke action will occur on the low tide. Fluke will sun themselves, warming up body temperatures and building appetites. Drifting these waters with a live killy and squid strip might be just the ticket for satisfying these appetites. (If it is still legal when you read this book, a strip of fluke belly might make even better "white bait.")

Best results in the western portion of the bay are rung up in the main east-west channel. In *Good Fishing Close to New York City*, author Jim Capossela suggests trying the cut north of buoy 22, just on the west side of the Ponquogue Bridge. Another spot to find flounder and fluke is the southeast corner of Tiana Bay, at buoy 19.

Shinnecock Bay is an especially good place to bring the kids. As Manny points out in his book *Gone Fishin' with Kids* (published by Gone Fishin' Enterprises, Annandale, New Jersey), leave your own fishing tackle at home if you take as many as two kids fishing. If you are fishing seriously, the children will not have fun. And fun is the name of the game if you want to get them hooked on a sport you might someday love together.

95 Shinnecock Canal

Directions: Follow the Sunrise Highway to the town of Hampton Bays and the canal.

The Shinnecock Canal was dug in the late 1800s to provide access from Shinnecock Bay to Great Peconic Bay. The opening was improved after World War II with the building of jetties.

True to the Shinnecock name, the canal is an outstanding place to fish, especially for flounder, fluke, and weakfish. However, as author Nick Karas points out in his book *Guide to Saltwater Fishing in New York*, when the tide is running and the locks are open, "there is no way to fish because of the current's intensity." Thus the time to fish the canal is when the gates are closed. (Knowing when the gates close is not an exact science;

generally speaking, however, the gates close approximately two hours before high tide, since they close automatically when the tide runs hard enough. We recommend you read the tide chart in your daily newspaper or *The Fisherman* magazine, look for Shinnecock Inlet and add two hours.)

The south side of the canal is usually quietest, harboring fish that stay inside through openings and closings of the gates, as well as fish that swarm in with each new opening. (The hottest action often takes place just before the gates open again.)

Even though flounder are lethargic in early spring, the combination of slow water inside the canal and a sunny day in March often will trigger feeding.

In the fall, flounder fresh in from the ocean often feed ravenously during their brief stays in the canal. Many stay only through one tide change until the canal reopens and they can head into Shinnecock Bay to fatten up before digging in for the long, hard winter ahead.

In summer, the canal is a first-rate location for fluke. When you find fish, remember that they are not likely to hold in one place for very long. Tides cause them to move, as does the opening of the gates; thus it is important to work quickly, because you will not have many hours to fish. We recommend a modified high-low rig for fluke. Instead of a typical long leader below a three-way swivel, change to a much shorter leader no more than 24 inches long below your weight. Above your sinker knot you may want to tie in a dropper loop and add a 10-inch leader with another hook, about 2 feet up the line. Bait the bottom hook with a strip of squid and a live killy, and the top hook with one or the other, but not both, since you want that hook to swing more freely than the bottom hook.

Weakfish also trade back and forth in the canal and shedder crab bait or 3-inch-long, .5-inch-wide tapered squid strips often induce them to bite. Many weaks are lost, however, when hooks tear out of their soft mouths on the lift. You can increase your odds of landing your fish with the use of a hook that features a thicker-than-usual shank. We like a size #2 bronze beak-style Eagle Claw or Mustad hook with baitholder to handle this job.

The Shinnecock Canal obviously necessitates a bit of additional thought, especially in determining when to fish. However, the results more than justify the effort.

96 Shinnecock Inlet

Directions: Take Dune Road to the inlet, located between Tiana Beach and Southampton Beach off the tip of Southampton Bays.

Aside from Montauk, the word Shinnecock is most identified with superb saltwater fishing in New York. Whether fishing the Shinnecock Inlet, the Shinnecock Bay, Shinnecock Canal, or the waters outside the inlet—all discussed in this book—Shinnecock is an area close to the hearts of Empire State anglers who fish the brine.

Shinnecock Inlet is a virtual fish highway between Shinnecock Bay and the Atlantic. A truly remarkable volume of striped bass feed in and around the inlet, especially at night.

Catching big bass here does not require a boat. Fishing from jetties or the rocks can be very productive. Experienced anglers cast rubber eels, bucktails, or Hopkins, the old reliable striper lure. The key is to vary the speed of the retrieve. Most anglers feel that a slow but staccato reeling action is best, working the lure in a stop-and-go fashion. (If fishing from the often-slippery rocks, wear appropriate footgear to prevent a nasty fall.)

As an aside, Captain Ray Kelly of TFN (The Fishing Network) believes that stripers here like to eat small flounders, a quite natural food fish. Since it is illegal to use undersized flatties, a lure like the Peri-Fish fake flounder is a good substitute. (In fact, all large-mouthed fish will eat small flatfish, as evidenced by a codfish Manny caught in Quincy Harbor, Massachusetts, which had seventeen little flounders and one baby fluke in its belly.)

The west bank of the inlet is within Shinnecock Inlet County Park and provides excellent fishing access. (At this writing, there were no residency requirements to fish in the park.)

Each summer, fluke are commonly caught in the inlet, and again in fall, as the big flatfish head out to sea. Loads of blackfish are caught just west of the jetty, and weakfish, running to considerable size, occasionally trade back and forth in the inlet between bays. Many more bluefish are caught in the inlet, often by anglers who are rigged and ready when a school of frantic baitfish starts bouncing up and out of the water close to the rocks. Having a second rod at hand rigged with a shiny lure is always recommended for these moments.

Aside from surf-casting spots, Shinnecock Inlet is one of the few saltwater locations where shoreline angling is the special focus. It is quite feasible to fish in the inlet from a boat, especially for fluke on the tide changes. But the tide can be furious here and small boats can be at risk in all but the slackest water. We urge you to exercise caution and good common sense.

97 Shinnecock Inlet, Outside

Directions: Sail out of Shinnecock Inlet between 1 and 2 miles to waters between 30 and 40 feet deep.

The waters outside of Shinnecock Inlet are prime fishing grounds for striped bass and fluke, especially favored by Captain Judy Schoerlin, who operates her 17-foot center console, *Hey Jude*, out of Molnar's Landing in Hampton Bays. Judy points out that one doesn't need a very large craft in order to fish "outside." (Indeed, our own vessels, a Sea Nymph and a Boston Whaler, both are smaller than the good ship *Hey Jude*, and we have ventured out more than a few times into nasty winds!) We just make sure, as you should, that when moving outside, we know the predicted wind direction; wind cranking in from the sea can make things dangerous for a small boat, and almost guarantees a sloppy, uncomfortable ride.

Judy fishes anywhere from 1 to 2 miles off the beach, depending upon water temperature and clarity. If the water is clear and warm, the fishing is better closer in. However, if the wind has been blowing from offshore

Figure 33 ■ Judith Schoerlin took this big "flattie" off of Hampton Bays. *(Photo: Judith Schoerlin.)*

for a few days, the water is generally a bit dirty; hence the fishing is better further out, closer to 2 miles out than 1, where the water will be cleaner.

Getting there is the easy part. Sail out of Shinnecock Bay and through the inlet and you are "at sea." Then either hang a left and head east toward Southampton Beach or turn right and go westerly in the direction of Quogue. To locate good places to stop your engine and start your drift, look for several prominent landmarks. If you go east out of the inlet, look for a group of pyramid-shaped and castle-shaped houses at Southampton. If you go west, look for the Ponquogue Bridge, which is inside the bay and a little to the west off of Hampton Bays. Mark these landmarks to start your drifts up and down the beach. A southwest wind is the very best you can get.

The rig Judy recommends for fluke is a 3-ounce sinker on a three-way swivel with the hook a size 3/0 in wide gap and a small chartreuse twister tail at the eye of the hook for added attraction. The hook bait should be a strip of squid and a spearing combination.

A ball-shaped white bucktail with white twister tail is another good choice, especially when the tide and wind are modest. In quiet water, you provide the action by lifting the jig off bottom a few inches, then varying the lift by slamming the jig upward from time to time.

Judy also scores well in these same waters on striped bass. In fact, she caught a 33-pounder just outside the inlet on a 5/0 hook with fish-finder rig and 2-ounce sinker. For really big bass, she recommends live eels.

The action outside the inlet is best from June to October, especially since a nice weakfish or two often will spice up the catch.

98 Peconic Bay

Directions: Take Route 27 east to the bay, or Route 24 north to the top of the bay at Riverhead.

Peconic Bay is actually a series of bays and channels, all loaded with fish and most especially with weakfish (sea trout). These waters have produced more "weaks" than virtually any other site in the area, and not just in New York. For example, the Mullica River and Delaware Bay in New Jersey are undoubtedly weakfish havens. But year in and year out, Peconic Bay produces weaks at an even higher level.

The style of fishing has remained pretty constant over the many years since we first fished the bay back in the 1950s. A heavy spinning rod or a good bait-casting stick is recommended. A high-low rig, with one snelled hook dropping your #2 baitholder hook to a point a foot above the sinker and another tied in above the sinker knot is all the rig you need. The preferred bait for weakfish is a piece of shedder crab. Sandworm also is high on the menu, as are live grass shrimp.

Little Peconic and Great Peconic Bays both contain porgies as well as weakfish, with weaks especially concentrated south of Robins Island. Toward the westerly end of Great Peconic, the fishing is first rate for winter flounder before and after the porgy and sea trout runs.

The flatties bite best in March, April, and November. During the other months, count on porgies, from hand-sized to 2-pounders, and weakies ranging upward in size from spikes to double-figure-sized beasts.

Peconic Bay also is known as a super fluke fishery. Gigantic summer flounder start showing up in the bay in May. A few years ago, a Peconic Bay "dormat" weighing in excess of 15 pounds was taken on board a charter boat called the *Orient Star*.

The top bait for fluke here is a strip of squid combined with a sand eel. A live killy with a squid strip is almost as effective. Most important, fish a moving tide, dragging your bait just at bottom. In slow tides, make sure to use live bait. (Dead fish such as spearing or sand eels work when the tide is moving fastest.) An excellent area for fluke is off Greenport in Little Peconic Bay.

99 Orient Point

Directions: Take Route 25 to Orient Beach State Park.

Orient Point is called "the tip of the North Fork" for good reason. It is as far northeast as one can go on the north shore. Essentially, Orient Point comprises several totally different fisheries, from land and from sea. Fishing from shore, anglers cast into the northeast tip of Long Island Sound or into quieter waters facing east. To the south, anglers fish from shore off Greenport. Boat anglers head out of three sites in Greenport. They are at Third and Wiggins Streets, at the Stirling Basin Launch Ramp on Sandy Beach Road, or at Norman Klipp Marine Park on Manhasset Avenue.

When the winds and weather are right, all may enjoy fine catches of flounder. Boats heading below the fork enjoy especially good flounder action in the spring and fall. While conventional wisdom says fish for flatties in March and April, some good catches are registered as the weather cools and the fish return to spawn and hibernate. The fish taken in fall are actually thicker and tastier, conventional wisdom notwithstanding.

The channel between Orient and Plum Island offers wonderful striped bass and bluefishing. Remember, though, that you are very close here to wide-open water. Make sure your motor is running true and smooth. An outgoing tide with a dead kicker could really put a dent in your day, or worse!

Look for bird action in this area. When you see diving gulls, slowly move toward the action. Have a spinning outfit at the ready with an Ava-style jig attached, ready to cast for blues. (In fact, it is best to have two rods available, one with the plain jig and another with a green or red tube attached.) Rigging a size #2 plain white bucktail hook 2 feet above the jig may net a double header when the action is hot.

To the northeast of the point, there is a 7-mile length of shoreline running to Truman Beach that offers good action from shore for fluke. At the lighthouse, you can catch blackfish on crab bait around the rocks. (Wear creepers to prevent a nasty fall on the slippery rocks at low tide.)

Orient Point provides a veritable potpourri of action, all within a limited space, whether fished from boat or shore.

100 Montauk Rips

Directions: The "rips" can be found 1 mile offshore between Montauk Harbor and Montauk Light.

A "rip," simply put, is a severe change in water flow; a turbulent place in the water created by the push and pull of tides. It is a more or less permanent phenomenon, often occurring in otherwise calm water. "The Race" off the coast of Connecticut is one of the better known rips in our region, a place where the current is so violent that anglers fish with heavy "drail"-type sinkers: rigs with chains at one end and monster hooks at the other.

The primary quarry at the Race is big bluefish. However, in the Montauk Rips, the name of the game is fluke. Baitfish get trapped in the swirling waters, attracting big, opportunistic fluke in great numbers.

You won't need "drail" sinkers here, but you will need heavy spinning tackle and sufficient lead, even if you are catching fish averaging only a pound or two. Thirty-pound test line is fine, with some anglers swearing by the nonstretch line to feel the bite better in a thrashing sea. Light tackle simply won't handle the current. The gear and line made available on a headboat or charter boat will be well suited to the task.

If you are using a single hook, it should be gold in color and a size ranging from 2/0 up to 4/0. We like to use an English bend (wide gap) hook of extremely small size inshore, as small as a size 4, and plain monofilament as leader. But we go with bigger hooks here because the fish are larger and the tides require stiffer leader.

If you plan to take your own boat to the Rips or to rent a boat at Montauk, be sure the winds are light. You don't want to get stuck out there in high winds, even if the distance back seems short. In fact, a mile can seem an impossible distance when banging back into headwinds of 20 miles an hour or more. Actually, the best approach to catching fluke in the Rips is to fish one of the many charter or headboats out of Montauk. The skippers know what they are doing, and they will bring you to the fish.

Basic baits are a "sandwich" of squid and spearing or fillet of sea robin and sand eel. The key is a long, tapered fillet cut out of a fresh fish, whether a fluke (if legal), a sand shark, a sea robin, or even a small bluefish. We always take mackerel or herring fillets that we keep frozen at home for such occasions, too. An unadorned smelt also can do the job. The key to fishing the Rips for big fluke is the familiar equation: "Big bait equals big fish."

There was a time when the Montauk Rips yielded big catches of codfish, and we look forward to their return. However, in the meantime, we're perfectly happy to fish for big fluke in these interesting, roiling waters.

Index

Numbers in **boldface** refer to key maps, pp. xxii–xxv

About the Authors

Ron Bern is the author of four books, including *The Legacy*, a novel. **Manny Luftglass** writes a regular column for *The Fisherman* and is the author of five other books about fishing. Together, they co-authored *Gone Fishin': The 100 Best Spots in New Jersey* (Rutgers University Press, 1998).